Astrophysics at Ultra-High Energies

THE SCIENCE AND CULTURE SERIES – ASTROPHYSICS

Series Editor: A. Zichichi

International School of Cosmic Ray Astrophysics
15th Course

Astrophysics at Ultra-High Energies

Erice, Italy 20 – 27 June 2006

Edited by

Maurice M Shapiro
University of Maryland, USA

Todor Stanev
University of Delaware, USA

John P Wefel
Louisiana State University, USA

World Scientific

NEW JERSEY • LONDON • SINGAPORE • BEIJING • SHANGHAI • HONG KONG • TAIPEI • CHENNAI

Published by

World Scientific Publishing Co. Pte. Ltd.
5 Toh Tuck Link, Singapore 596224
USA office: 27 Warren Street, Suite 401-402, Hackensack, NJ 07601
UK office: 57 Shelton Street, Covent Garden, London WC2H 9HE

British Library Cataloguing-in-Publication Data
A catalogue record for this book is available from the British Library.

International School of Cosmic Ray Astrophysics
ASTROPHYSICS AT ULTRA-HIGH ENERGIES – 15th COURSE

ISBN-13 978-981-279-014-9
ISBN-10 981-279-014-4

Printed in Singapore by World Scientific Printers (S) Pte Ltd

PREFACE

It is becoming increasingly clear that we live in a High Energy Universe with the acceleration of particles to Ultra-high Energy (UHE) as the underlying cause. These particles interact to produce Gamma-rays and Neutrinos as well as surviving to be observed a Ultra-high Energy Cosmic Rays. Under the auspices of the International School of Cosmic Ray Astrophysics (M. M. Shapiro, Director), this 15th biennial Course entitled "Astrophysics at Ultra-high Energies" brought together students, faculty and researchers to explore the exciting new work that is underway at UHE. The school featured a full program of lectures and discussion in the ambiance of the Ettore Majorana Centre in Erice, Italy, including visits to the local Dirac and Chalonge museum collections as well as a view of the cultural heritage of southern Sicily. This course was attended by 60 participants from 15 different countries.

The program provided a rich experience, both introductory and advanced, to the inter-connected areas of High Energy Astrophysics: powerful astrophysical sources, ultrahigh energy cosmic rays, gamma ray astronomy and ultra-high energy neutrinos. Gamma ray bursts, as observed on the SWIFT Spacecraft, were described and possible sources, most involving massive black holes, were analyzed. In the TeV region, atmospheric Cherenkov telescopes have matured into a new observation tool that can study a large variety of high energy source objects. New technical advances in gamma-ray astronomy are underway making this an important area for future discovery. Moreover, "neutrino astronomy" is on the verge of becoming a new window to the univese and the techniques, instruments under development, preliminary results and the anticipated sources and propagation of these particles were addressed by experts in the field. Finally, recent advances both on the experimental side and in the theory and interpretation of UHE cosmic ray particles were fully discussed.

Contained in this volume is a collection of the lectures and presentations made at the School involving the physics and astrophysics of the newly emerging research area that already has been, and will continue to be, as important contributor to understanding our high energy universe. The volume is suitable for students and advanced researchers wanting a current picture of the high energy situation, both experiment and theory, either for personal use or as part of a course of study for advanced students.

A highly successful course requires the combined effort of many individuals, foremost of whom are the Lecturers who give their time and expertise in both formal presentations and informal discussions. To them goes our heartfelt thanks. This course was co-directed by J. P. Wefel and T. Stanev. We acknowledge G. Sutton, and S. Rowland-Perry for their help with the organization, program and manuscripts. Executive Secretary A. Smith helped keep everything running efficiently.

We were delighted with the exceptional facilities of the Ettore Majorana Centre for Scientific Culture, the host for this school, and we acknowledge Centre Director A. Zichichi, plus Fiorella, Pino, Alessandro, Alberto, and a host of others who contributed to the course. We also thank the Sicilian Regional Government, the Italian Ministry of Education, and all of the institutes, universities and government agencies who helped to support the participants — the true beneficiaries of the Course.

M. M. Shapiro, T. S. Stanev and J. P. Wefel

CONTENTS

Gamma Ray and Neutrino Astronomy

powerful astrophysical sources

GAMMA-RAY BURST: DISCOVERIES WITH *Swift*

ALAN WELLS

Space Research Centre, Department of Physics and Astronomy, University of Leicester, Leicester, LE1 7RH, UK

Gamma Ray Bursts (GRBs) are bright, brief flashes of high energy photons and are the most powerful explosions since the Big Bang, with typical energies up to around 1051 ergs. Their outbursts persist for durations ranging from milliseconds to tens of seconds or more. In these brief moments the explosions radiate more energy than the Sun will release in its entire 10 billion year lifetime. They come in two classes: long (¿2 s), softspectrum bursts and short, hard events. Current theories attribute these phenomena to the final collapse of a massive star, or the coalescence of a binary system induced by gravity wave emission.

New results from Swift and related programmes offer fresh understanding of the physics of gamma-ray bursts and of the local environments and host galaxies of burst progenitors. Bursts found at very high red-shifts are new tools for exploring the intergalactic medium, the first stars and the earliest stages of galaxy formation.

1. Introduction

Gamma Ray Bursts (GRBs) were first discovered in the late 1960s (Klebesadel et al. 1973). They come in two classes: long (¿2 s), soft spectrum bursts and short, hard events (Kouveliotou et al. 1993). Results from the Compton Gamma-ray Observatory (CGRO) showed them to be distributed isotropically over the sky occurring at a rate of about 300 per year (Meegan et al. 1991). The BeppoSAX mission made the important discovery of X-ray afterglows associated with long bursts. (Costa et al. 1997). Follow-up observations found afterglows at optical (van Paradijs et al. 1997) and radio (Frail et al, 1997) wavelengths and provided redshift (and hence distance) measurements to place upper bounds on the total energy (¿1051 ergs/s) of the bursts. Identification of the hosts showed-at least for the Deposal sample-that long GRBs emanated from regions of high star formation rate in high redshift galaxies. The afterglow observations provided compelling evidence in support of the fireball model which associates the burst and the subsequent afterglow with shocks generated within highly relativistic jets ejected from the progenitor (Meszaros & Rees, 1997). Deposal and HETE-2 between them found a small sample of closer GRBs, most notably GRB030329/SN2003d, in association with Type 1c supernovae pointing to collapse of the central core of a massive early type star and formation of a black hole as the precursor to the GRB outburst. (MacFadyen & Woosley. 1999). However, prior to Swift, most afterglow data were collected hours after the burst so little was known about the origins of the short bursts or about the early emission behaviour of the high red-shift long bursts.

2. Observations with *Swift*

The Swift satellite (Gehrels et al. 2004) was specifically designed to study early GRB emissions and to detect the afterglows by automatically slewing to a GRB as soon as it had been detected on-board. Swift carries a sensitive coded mask Burst Alert Telescope (BAT) and finds new GRBs by detecting gamma ray emission in the 15-150 keV range. When BAT detects a new burst, subject to certain visibility constraints, Swift autonomously re-points to bring the GRB within the field of view of the X-ray Telescope and the UV/Optical Telescope. Observations with these instruments start very quickly after the initial burst detection from which precision location of the bursts, to arc-second accuracy, are usually obtained. Accordingly, Swift routinely provides prompt detections of GRBs and their afterglows and automatically transmits their locations and other information obtained from the three instruments via the TDRSS satellites and the Gamma-ray burst Coordinate Network (GCN) to observers and robotic telescopes around the world.

Swift was launched on November 20 2004 and has since been detecting GRBs at the predicted rate of 100 per year. At the time of the Erice meeting, 140 GRBs had been detected and, in most cases, the spacecraft was able to slew to the source within 5 minutes of the initial detection-often much more quickly. X-ray afterglows were detected on-board in virtually all of these promptly observed bursts, whereas optical/uv afterglows were detected on-board in only 30

Swift is locating many more high redshift (z¿2) bursts than previous missions. Indeed, the majority of Swift GRB detections to date have been of the long burst variety, and studies of the early afterglows, previously inaccessible, have added to evidence supporting the view that long duration bursts are produced during the collapse of a massive star. Redshifts have now been measured for over 50 long bursts including the first GRB at very high redshift (z¿6). (Cusumano et al., 2006, Haislip et al. 2006, Kawai et al. 2006.). These burst are providing new ways to probe the high redshift universe (Lamb, 2007 and Ghirlanda, 2007) and Tanvir & Jakobsson, 2007, discuss conditions under which GRBs may be used as a tracer of the star formation rate in high redshift galaxies.

Swifts multi-wavelength measurements (gamma-ray to optical/uv) of the exceptional nearby burst GRB 060218 (z = 0.033) have provided a direct observation of the shock breakout in a supernova collapse (Blustin, 2007, Campana et al. 2006, Zhang et al. 2007), this observation adding to the small sample of previously observed nearby long GRBs associated with supernova collapse. More recently, Swift has found two bursts, GRB 060614 and GRB060505 for which no supernova association has been found down to deep limits (Watson et al, 2007 and references therein).

Swift also made the first X-ray afterglow localisation of a short burst and has since found several more along with two additional detections with HETE-2. (Gehrels, 2007, and references therein). Most Swift short bursts have X-ray afterglow detections and about half have host identifications and redshifts. Gehrels, (2007), and Barthelmy, et al. (2007) remark on the similarity of the emission of GRB 060614 with the peak luminosity and spectral lags seen in short bursts whilst its afterglow emissions are more characteristic of a low-redshift long bursts despite the non-detection of a coincident supernova to deep limits.

Various new features of GRB phenomenology, such as the soft tail in the spectra of short hard bursts; X-ray flares in the early afterglow indicating extended activity of the central engine in GRBs; GRBs associated with supernovae; GRBs with no supernovae; energetic supernovae with no GRBs; are all discoveries from the Swift era. All offer new challenges to current theoretical understanding. Many have been addressed in the papers in this volume.

3. Models, progenitors and jets

Woosley and Zhang (2007) remark on this diversity of burst phenomena and argue in favour of a single basic model for the central engine operating in a massive star but allowing for variable pre-supernova mass and different rotation and mass loss rates. Metallicity is a key factor affecting all three of these properties. The central engine must generate both a narrowly collimated, highly relativistic jet to make the GRB, and a wide angle, sub-relativistic outflow responsible for exploding the star and illuminating the supernova. As the two components may vary independently, it is possible to produce a variety of jet energies and supernova luminosities. They go on to explore the production of low energy bursts and find a lower limit (1048 ergs/s) to the power that a jet requires in order to escape a massive star before that star either explodes or the core is accreted. Lower energy bursts may be particularly prevalent when the metallicity is high, i.e., in the nearby universe at low redshift. Conversely, Podsiadlowski (2007) discusses the potential of a variety of binary merger models to account for the diversity of long duration GRBs.

Pe'er et al (2007) suggest that thermal radiation may accompany the first stages of a GRB, to explain observed features in the prompt gamma emission which are inconsistent with the optically thin synchrotron emission more commonly associated with the fireball model. Lazzati et al. (2007) model hydrodynamic propagation of a relativistic jet through a massive star and find radiative phases in the jet propagation which could contribute to the GRB light curves. The scenario of jet evolution described in this work may also provide an explanation for the long dead-times between precursor and the main GRB emission seen occasionally with BAT and previously with BATSE bursts from CGRO.

4. Afterglows

Swift has filled the temporal gap between the prompt emission and the afterglow that earlier missions were generally unable to probe. O'Brien & Willingale (2007) have shown that light curves combined from BAT and XRT show an essentially smooth transition between the non-thermal prompt X-ray emission and the decaying X-ray afterglow. They and others (Piran & Fan, 2007, Panaitescu, 2007, Burrows et al. 2007 and references therein) agree on the generic nature of early GRB light curves, as illustrated in Figure 1, with the proviso that not all phases in the afterglow evolution shown in the figure are present in all bursts thus suggesting that several different dissipation processes may be involved.

When present as the dominant feature, the initial steep decay is usually attributed to large angle high latitude emission produced from internal shocks; when the slower unbroken power law decay is dominant this is attributed to forward shock emission from a narrow jet. Most bursts appear to exhibit a combination of both features. About half of the afterglows have X-ray flares superimposed on the broken power law curve, as illustrated in Fig. 1, (Burrows et al, 2007) also indicative of continuing activity within the central engine for extended periods after the initial outburst. Piran & Fan, 2007, Panaitescu, 2007, and others have discussed added complexity

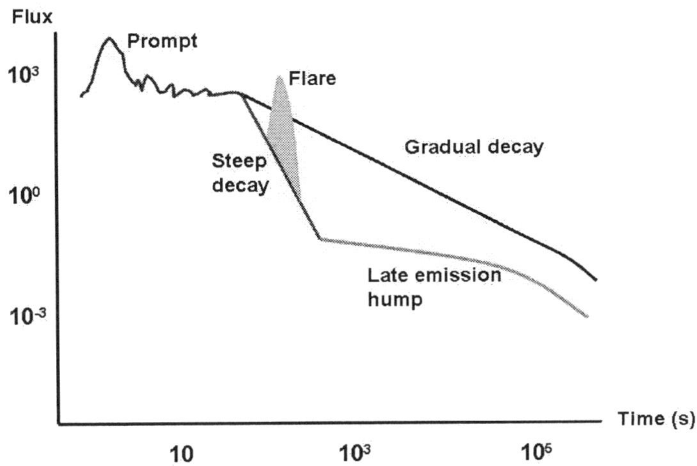

Fig. 1. A schematic view of the early GRB X-ray light curve. Following the prompt emission, which typically lasts a few 10s of seconds, the decay tends to follow one of two paths: (i) a steep decay, during which the flux can fall by several orders of magnitude, followed by a shallower, late emission hump starting at 10^3 s; or (ii) a more gradual decay. Either decay path can end with a break at $>10^4$ s to a steeper decay. X-ray flares can occur during either decay path, most prominently during the first hour. (O'Brien & Willingale, 2007)

to afterglow models needed to fully understand these new features.

Optical afterglows have been monitored on-board Swift with the UVOT telescope (Mason et al, 2007) as well as with ground based telescopes (e.g. Antonelli et al, 2007). Results indicate considerably more complexity in many bursts than would be expected from the standard fireball model and varying degrees of correspondence between the X-ray and optical light curves. In this respect GRB 050525A may be an exception as it exhibits an achromatic jet break at 104s, although, even in this case, the decay indices are shallow compared with what might be expected from the standard model (Zhang et al, 2007, and references therein). Afterglow behaviour from other bursts is not so easily interpreted and the absence of jet breaks in the X-ray afterglow, when present in the optical remains to be understood.

Radio monitoring of GRB afterglows, enable the evolution of GRB explosions to be monitored long after the X-ray and optical afterglows have faded. GRB 030329 is still visible at radio wavelengths 1100 days after the burst trigger (van der Horst et al., 2007) and continuing monitoring is revealing structural evolution of the burst with and indicating transition from a collimated relativistic outflow to spherical non-relativistic outflow during this period.

5. Short-hard gamma-ray bursts

Swift's discovery of the first afterglows from short-hard Gamma-ray bursts is being followed by a systematic study of short bursts through X-ray, optical and radio afterglow measurements of multiple short bursts. (Fox & Roming, 2007, Barthelmy, 2007, Levan, 2007, Zane, 2007, and references therein). Their distance scale (z ¿ 0.1) and energetics (E> 10^{48} erg) are now established, and they have been revealed definitively as a cosmological phenomenon. The short bursts have been found among old stellar populations in elliptical galaxies, galaxy clusters and the outskirts of younger galaxies and the absence of associated supernovae appear to rule out an origin in the deaths of massive stars. This is in contrast to the now accepted view of the origins of long-duration gamma-ray bursts (GRBs), whose host galaxies, redshifts, and associated supernovae are all consistent with the collapsar-supernova model.

The effect of these discoveries has been to strongly favour the compact-object merger model for short bursts. The observed properties point to coalescence of a compact-object binary, either neutron star-neutron star or neutron star-black hole (King, 2007) and enabling the prospects for gravitational-wave detection to be re-assessed (Hough, 2007).

Acknowledgments The author wishes to acknowledge his Leverhulme Emeritus Fellowship which has supported his recent work on Gamma ray bursts including participation in this meeting.

References

1. Antonelli, L.A., et al, 2007, Phil. Trans. R. Soc. A; Vol 365; in print.
2. Barthelmy, S.D. et al, 2007, Phil. Trans. R. Soc. A; Vol 365; in print.
3. Blustin, A.J., 2007, Phil. Trans. R. Soc. A; Vol 365; in print.
4. Burrows, D.N. et al., 2007, Phil. Trans. R. Soc. A; Vol 365; in print.
5. Campana, S. et al., 2006, A&A, 454, 113.
6. Costa, E. et al., 1997, Nature, 387, 783.
7. Cusumano, G., et al.,2006, Nature, 440, 164.
8. Fox, D.B. & Roming, P.W.A. 2007, Phil. Trans. R. Soc. A; Vol 365; in print.
9. Frail, D.A. et al., 1997, Nature, 389, 261.
10. Gehrels, N. 2007, Phil. Trans. R. Soc. A; Vol 365; in print.
11. Gehrels, N. et al., 2004, ApJ, 611, 1005.
12. Ghirlanda, G., 2007, Phil. Trans. R. Soc. A; Vol 365; in print.

13. Haislip, J.B. *et al.* 2006 Nature, 440, 181.
14. Hough, J., 2007, Phil. Trans. R. Soc. A; Vol 365; in print.
15. Kawai, N., et al. 2006, Nature, 440, 184.
16. King, A., 2007, Phil. Trans. R. Soc. A; Vol 365; in print.
17. Klebesadel, R.W. et al., 1973, ApJ, 182, L85.
18. Kouveliotou, C. et al., 1993, ApJ, 413, L101.
19. Lamb, D.Q., 2007, Phil. Trans. R. Soc. A; Vol 365; in print.
20. Lazzati, D. et al. 2007, Phil. Trans. R. Soc. A; Vol 365; in print.
21. Levan, A.J., 2007, Phil. Trans. R. Soc. A; Vol 365; in print.
22. MacFadyen, A.I. & Woosley, S.E. 1999, ApJ, 524, 262.
23. Mason, K.O. et al., 2007, Phil. Trans. R. Soc. A; Vol 365; in print.
24. Meegan, C.A. et al, 1991, Nature, 355, 143.
25. Mszros, P., & Rees, M.J., 1997, ApJ, 476, 232.
26. O'Brien, P.T. & Willingale, R. 2007, astro-ph/0701811
27. Panaitescu, A. 2007, Phil. Trans. R. Soc. A; Vol 365; in print.
28. Pe'er, A. et al., 2007, Phil. Trans. R. Soc. A; Vol 365; in print.
29. Piran, T., & Fan, Y-Z. 2007, Phil. Trans. R. Soc. A; Vol 365; in print.
30. Podsiadlowski, P., 2007, Phil. Trans. R. Soc. A; Vol 365; in print.
31. Tanvir, N.R. & Jakobsson, P. 2007, Phil. Trans. R. Soc. A; Vol 365; in print.
32. van der Horst, A.J. et al., 2007, Phil. Trans. R. Soc. A; Vol 365; in print.
33. van Paradijs, J. et al., 1997, Nature, 386, 686.
34. Watson, D., et al. 2007, Phil. Trans. R. Soc. A; Vol 365; in print.
35. Woosley, S.E. & Zhang, W., 2007, Phil. Trans. R. Soc. A; Vol 365; in print.
36. Zane, Z., 2007, Phil. Trans. R. Soc. A; Vol 365; in print.
37. Zhang, B. et al. 2007, Phil. Trans. R. Soc. A; Vol 365; in print.

GAMMA-RAY BURST PHENOMENOLOGY IN THE SWIFT ERA*

P. Mészáros

Dept. of Astronomy & Astrophysics and Dept. of Physics, Pennsylvania State University, University Park, PA 16802, USA
**E-mail: pmeszaros@astro.psu.edu*

Rapid follow-up of gamma-ray burst (GRB) afterglows with the multi-wavelength satellite *Swift* and other instruments is leading to a reappraisal and expansion of the standard model of the GRB early afterglow and prompt gamma-ray emission. The previously uncharted time range of minutes to hours has revealed systematic X-ray light curve properties such as steep decays, shallow decays and flares. Other discoveries include the localization and follow-up of short GRB afterglows, the detection of long bursts beyond the redfshift $z \gtrsim 6$, the detection of prompt optical/IR emission while the gamma-rays are still on, the detection and prompt follow-up of supernovae associated with GRB. We review some of the current theoretical issues.

Keywords: Gamma-ray bursts

1. Challenges posed by new Swift observations

NASA's *Swift* mission[43] has two new capabilities: a greater sensitivity of its Burst Alert Detector[126] (BAT; energy range 20–150 keV) compared to the preceding *Bepposax* and *HETE-2* missions;[42] and the ability to slew in less than 100 seconds to the burst direction determined by the BAT, allowing it to position its much higher-angular resolution X-ray (XRT, few-arcsec) and UV-Optical (UVOT, sub-arcsec) detectors[44,45] for observing the prompt and early afterglow emission.

Redshifts: The total number of redshifts since 1977 is now over 80. The $\gtrsim 60$ Swift-enabled redshifts have a median $z \sim 2.8$,[46,47] a factor $\gtrsim 2$ higher than those from previous satellites.[9] This is thanks to the prompt ~arcsec positions from XRT and UVOT, making possible rapid ground-based observations while the afterglow is still bright.

*To appear in Proceedings 2006 Erice Summer School of Cosmic Ray Astrophysics

BAT light curves: The BAT triggering algorithms are much more sophisticated than previously, including "imaging" triggers capable of detecting faint, slow-rising events. This made possible the discovery of both the farthest, highly time-dilated burst GRB 050904 ($z = 6.29$), and the nearest, faint and slow-rising burst GRB 060218. In some "long" bursts (duration $t_\gamma \gtrsim 2$ s), faint soft gamma-ray tails extend the duration by a factor up to two beyond previous BATSE values.[1] These have been found also in some "short" bursts (previously defined as $t_\gamma \lesssim 2$ s).

XRT light curves: New insights on the burst and afterglow physics have been forthcoming from detailed X-ray light curves, starting on average 100 seconds after the trigger. This suggests a canonical X-ray afterglow[16,17] with one or more of the following: 1) an initial steep decay $F_X \propto t^{-\alpha_1}$ with a temporal index $3 \lesssim \alpha_1 \lesssim 5$, and an energy spectrum $F_\nu \propto \nu^{-\beta_1}$ with energy spectral index $1 \lesssim \beta_1 \lesssim 2$, extending up to times $300\text{s} \lesssim t_1 \lesssim 500\text{s}$; 2) a flatter decay $F_X \propto t^{-\alpha_2}$ with $0.2 \lesssim \alpha_2 \lesssim 0.8$ and $0.7 \lesssim \beta_2 \lesssim 1.2$, at $10^3\text{s} \lesssim t_2 \lesssim 10^4\text{s}$; 3) a "normal" decay $F_X \propto t^{-\alpha_3}$ with $1.1 \lesssim \alpha_3 \lesssim 1.7$ and $0.7 \lesssim \beta_2 \lesssim 1.2$ (generally unchanged the previous stage), up to a time $t_3 \gtrsim 10^5$s ; 4) In some cases, a steeper decay $F_X \propto t^{-\alpha_4}$ with $2 \lesssim \alpha_4 \lesssim 3$, after $t_4 \sim 10^5$s; 5) In about half the afterglows, one or more X-ray flares, starting as early as 100 s and sometimes as late as 10^5s, whose energy is $\sim 0.01 - 1$ of the prompt emission. The rise and decay is sometimes unusually steep, depending on the reference time t_0, i.e. $(t - t_0)^{\pm \alpha_{fl}}$ with $3 \lesssim \alpha_{fl} \lesssim 6$.

Very high redshift bursts: A major advance from Swift was the discovery of long bursts at $z > 5$, with GRB 050904 breaking through the $z \sim 6$ threshold. This burst was very bright, $E_{\gamma,iso} \sim 10^{54}$ erg. Ground-based OIR photometric upper limits and a J-band detection suggested a $z > 6$,[8] while spectroscopic 8.2 m Subaru telescope observations gave $z = 6.29$.[10] The X-ray brightness of the afterglow exceeded for a day that of the most distant X-ray quasar, SDSS J0130+0524, by up to 10^5 in the first minutes.[11] It also showed a prompt, very bright IR flash,[12] comparable to the famous $m_V \sim 9$ optical flash in GRB 990123.

The GRB-SN connection: An exciting Swift result was the observation, with BAT, XRT and UVOT, of an unusually long ($\sim$ 2000 s), soft burst, GRB 060218,[49] associated with the nearby ($z = 0.033$) SN2006aj, a type Ic supernova.[50–52] This was the first time that a connected GRB/SN event was observed from the first $\sim$100 s in X-rays and UV/optical. The early X-ray spectrum is initially dominated by a power-law component, with an increasing black-body component which dominates after $\approx$3000 s. Another Swift-enabled GRB/SN detection is that of GRB 050525A.[53]

Short bursts: A major advance from Swift was the localization, for the first time, of short GRB afterglows. As of October 2006 fourteen short bursts had been localized by Swift, in nine of which an X-ray afterglow was measured, while eight showed an optical afterglow, and one had a radio afterglow. This allowed, for the first time, the identification of host galaxies; these are of early type (ellipticals) in roughly half the cases, and dwarf irregular in the rest – with evidence for old as well as young stellar populations.[54] The median z (except for a few which appear at $z \gtrsim 1.8$) is $z_{\rm med} = 0.26$, about 1/3 that of the long bursts. While there is star formation in roughly half the host galaxies, overall the host properties correspond to those of an old progenitor population. Most short bursts localized by Swift have relatively low γ-ray luminosity, e.g. $E_{iso} \sim 10^{48} - -10^{50}$ erg. The X-ray afterglows roughly resemble those of long GRB.

2. Prompt gamma-ray emission

Most of the prompt gamma-ray emission of GRB is in the 0.1 to 2.0 MeV range, the spectrum being a broken power law[55] (with a number of bursts showing a thermal component as well[39]). The progenitors of "long" bursts are located in active star-forming galaxies, and are thought to be stars of initial mass $\gtrsim 25 - 30\ M_\odot$, the collapse of whose central core leads to a black hole, or possibly to a temporarily stabilized over-massive neutron star. Liberation of gravitational energy in the ensuing accretion explains the energy and timescales.[128–130] A plausible progenitor for "short" burst are the merger of compact binaries (NS-NS, or NS-BH),[131–133] leading to a black hole with a shorter accretion timescale. This scenario is only now beginning to be observationally tested with Swift.

The MeV-GeV emission of GRB is generally understood in terms of leptonic processes in the standard GRB fireball shock model. This involves a relativistic fireball undergoing shocks where particles accelerate, e.g. by a Fermi mechanism. Electrons radiating via synchrotron and inverse Compton produce the MeV radiation, and later also the increasingly softer electromagnetic afterglow.[40,41] The high Lorentz factors and inverse Compton (IC) scattering of synchrotron photons leads to the expectation of GeV and TeV photons.[88] Alternatively, the γ-ray emission may arise in Poynting dominated jets[134] due to magnetic dissipation accelerating leptons which radiate by synchrotron and/or IC.[56]

In either Fermi or reconnection schemes, a number of effects can modify the simple synchrotron spectrum. For instance, the distribution of observed low energy spectral indices β_1 (where $F_\nu \propto \nu^{\beta_1}$ below the spectral peak) has a mean value $\beta_1 \sim 0$, but for a fraction of bursts this slope reaches positive values $\beta_1 > 1/3$ which are incompatible with a simple synchrotron interpretation.[57] Possible explanations include synchrotron self-absorption in the X-ray[58] or in the optical range up-scattered to X-rays,[59] low-pitch angle scattering or jitter radiation,[60,61] observational selection biases[62] and/or time-dependent acceleration and radiation.[63] Other models invoke a photospheric component and pair formation.[64] A moderate scattering depth can lead to a Compton equilibrium which gives spectral peaks in the right energy range,[65,66] a high radiative efficiency, and spectra with steep low energy slopes.[67–69] It can also explain the Amati[104] or Ghirlanda[70] relations between spectral peak energy and burst fluence.[67,71]

3. Models of early afterglows in the Swift Era

The afterglow becomes important after a time $t_{ag} = \mathrm{Max}[t_{dec}\ , T]$ where the deceleration time is $t_{dec} \sim (3/4)(r_{dec}/2c\Gamma^2) = 10^2(E_{52}/n_0)^{1/3}\Gamma_2^{-8/3}(1+z)$s and T is the duration of the prompt outflow, t_{ag} marking the beginning of the self-similar blast wave regime. Denoting the afterglow spectral energy flux as $F_\nu(t) \propto \nu^{-\beta} t^{-\alpha}$, the late X-ray afterglow phases (3) and (4) described above are similar to those known previously from Beppo-SAX. (e.g.[41]). The "normal" decay phase (3), with temporal decay indices $\alpha \sim 1.1 - 1.5$ and spectral energy indices $\beta \sim 0.7 - 1.0$, is what is expected from the forward shock late time regime, under the assumption of synchrotron emission. The late steep decay decay phase (4) of §1, occasionally seen in Swift bursts, is generally explained as a jet break, when the decrease of the ejecta Lorentz factor leads to the light-cone angle becoming larger than the jet angular extent, $\Gamma_j(t) \gtrsim 1/\theta_j$ (e.g.[41]). However, this final steepening has been seen in $\lesssim$10% of the Swift afterglows, mainly in X-rays. The corresponding optical breaks have been few, and not well constrained. This is unlike the case with the $\sim$ 20 Beppo-SAX bursts, for which an achromatic break was reported in the optical,[13] while in rare cases where an X-ray or radio break was reported it occurred at a different time.[14] The relative paucity of optical breaks in Swift afterglows may be an observational selection effect due to the larger median redshift.

3.1. *Prompt optical emission*

Prompt optical flashes are defined as those detected while the gamma-ray emission is still ongoing, e.g. GRB 990123.[72] They are generally interpreted[73–75] as being due to the reverse external shock, although in principle they can arise from either an internal shock or the reverse external shock.[37,74] The optical flux decay rate in both cases is very fast, compared to the forward shock, so the latter typically dominates after tens of minutes. Prior to Swift, prompt and also early (those occurring within minutes but after the γ-ray emission ceased) optical flashes were rare, but are now detected with the Swift UVOT telescope at early times, $\lesssim$100 s, in roughly half the bursts.[48] A new discovery from robotic ground-based telescopes in the Swift era has been a gamma-ray correlated component of the prompt optical emission e.g. in GRB 041219.[76–78] This correlated component, while not observed in every burst, is suggestive of an origin in internal shocks.[37,74]

In contrast to bursts with reverse-shock flash or gamma-ray correlated emission, the typical bursts tend to show either a single power-law decay from early times,[79–81] or a flat or rising light curve[82,83] before entering a standard power-law afterglow decay. The initial brightness is typically $V \sim 14$ to 17 mag, which has made observations challenging for the usual $<$1 m telescopes employed. There are a number of possible reasons for the faintness of the early optical emission in typical GRBs. Suppression of internal shock emission can be due to self-absorption in the optical and/or the lower flux from the $\nu^{1/3}$ low-energy synchrotron asymptote.[37] Reverse shock emission suppression may also occurr if the ejecta are highly magnetized,[37] or deceleration might occur in the thick-shell regime ($T \gg t_{dec}$), so the reverse shock is relativistic and boosts the optical spectrum into the UV.[84] Pair formation in the ejecta could cause the reverse shock spectrum to peak in the IR.[85] More generically, accurate calculations of the reverse shock[86,87] find the emission to be significantly weaker than estimated earlier. Another possibility is that the cooling frequency in the reverse shock is not much larger than the optical band frequency, so after the reverse shock crosses the ejecta there are no electrons left to radiate in the optical.[87]

There have been a few convincing reverse-shock type optical flashes in the *Swift* era. E.g. GRB 041219 would have rivalled GRB 990123, except for the large Galactic extinction,[76] and one of its three peaks observed with PAIRITEL may be a reverse shock.[77] Observations of GRB 050525A with UVOT[89] and GRB 060111B with TAROT[90]) show the "flattening" lightcurve familiar from GRB 021211, which is termed a "type II" light curve by,[38] ascribed to a magnetized ejecta. GRB 060117, observed by the FRAM telescope of the Auger Observatory[91] had a bright optical flash peaking at $R \approx 10$ mag, which looks like a forward shock peak distinguished above the decay of the reverse shock flux, as in a "type I" two-peaked lightcurve from a low magnetization flow.[38] The most exciting prompt robotic IR detection (and optical non-detection) was GRB 050904,[8,12] also thought to be due to a reverse shock[92,93] (c.f.[94]). It was at an unprecedented $z = 6.29$,[10] witn an O/NIR brightness in the first 500 s (observer time) comparable to that of GRB 990123, and a steep time-decay slope $\alpha \sim 3$.

3.2. *Steep X-ray decay*

Among the new early afterglow features detected by Swift, the ubiquitous steep initial X-ray decay $F_\nu \propto t^{-3} - t^{-5}$ is one of the most puzzling. The most widely considered explanation for this fast decay, in the initial phase (1) and in the steep flares, is the off-axis emission at $\theta > \Gamma^{-1}$ (curvature, or high latitude emission[15]). After the line of sight gamma-rays have ceased, the emission from $\theta > \Gamma^{-1}$ is $(\Gamma\theta)^{-6}$ weaker than that from the line of sight. Integrating over the equal arrival time region, the flux ratio is $\propto (\Gamma\theta)^{-4}$. Since the emission from θ arrives $(\Gamma\theta)^2$ later than from $\theta = 0$, the flux is seen falling as $F_\nu \propto t^{-2}$, if the flux is frequency independent, while for a source-frame flux $\propto \nu'^{-\beta}$, the observed flux varies as $F_\nu \propto (t - t_0)^{-2-\beta}$, i.e. $\alpha = 2 + \beta$. This "high latitude" radiation appears to arrive delayed by $t \sim r\theta^2/2c)$ relative to the trigger time, and its spectrum is softened by the Doppler factor $\propto t^{-1}$ into the X-ray observer band. The flux is measured as $F_\nu \propto (t - t_0)^{-2-\beta}$, where t_0 can be the trigger time, or a value which is a fraction of the deceleration time emission can have an admixture of high latitude and afterglow.[22] Values of t_0 closer to the onset of the decay lead to steeper slopes. For the flares, if they are due to late internal shock or dissipation, the value of t_0 is just before the flare.[16] This appears to be compatible with most Swift afterglows.[16–18]

3.3. *Shallow X-ray decay*

The shallow decay portion of the X-ray light curves ($\alpha \sim -0.3 - 0.7$) is not entirely new, having served as motivation for the "refreshed shock" scenario,[20,21] which can flatten the afterglow light curve for hours or days. This occurs even even if the ejecta mass is disgorged promptly during $t = T \lesssim t_\gamma \sim 10 - 100$, s but with a range of Lorentz factors, say $M(\Gamma) \propto \Gamma^{-s}$, so that the lower Γ shells arrive much later to the foremost fast shells which have already been decelerated. Thus, it is not necessarily the central engine which is active late, but its effects are seen late. For fits of refreshed shocks to observed shallow decay phases in Swift bursts see.[24] Alternatively, one can envisage a central engine activity extending for long periods of time, e.g. $\lesssim$ day (in contrast to the $\lesssim$ minutes engine activity just mentioned abov e). This may be due to continued fall-back onto the central black hole,[28] or due to a magnetar wind.[23] The refreshed shock model can generally explain the flatter temporal X-ray slopes from Swift. However, the fluence ratio in the shallow X-ray afterglow and the prompt gamma-rays can reach $\lesssim 1$.[22] This requires a higher radiative efficiency in the prompt emission than in the afterglow. This might be achieved if the prompt outflow is Poynting-dominated, or if the afterglow emits more energy in other bands, e.g. GeV, or IR. Or[26] a previous mass ejection might have emptied a cavity, leading to an energy fraction going into the electrons $\propto t^{1/2}$.

3.4. *X-ray flares*

A possible explanation for X-ray flares which are not too steep in time is through refreshed shocks. This does not work for flares with time indices $\alpha \sim \pm 5 - 7$, such as in GRB 0500502b,[25] where also the flare flux level is a factor ~ 500 above the afterglow baseline. Inverse Compton scattering in the reverse shock[19] could explain a single flare at the beginning of the afterglow, wich is not too steep. For multiple flares, models invoking encountering a lumpy external medium have generic difficulties explaining steep rises and decays,[16] although extremely dense, sharp-edged lumps, if they exist, might satisfy the steepness.[27]

A widely considered model for the flares ascribes them to late central engine activity.[16–18] These can explain the fast rise and decay, and are easier on the energy budget, e.g. the flare energy can be comparable to the prompt emission. The fast rise comes from the short dissipation time, and the rapid decay is due to high latitude emission with t_0 reset to the beginning of each flare.[16] A central engine origin is conceivable, within certain time ranges, based on numerical models of core collapse long bursts.[28] However flare fluences which are a sizable fraction of the prompt emission hours later are difficult to understand. Flaring may also be due to gravitational instabilities in the infalling debris,[29] or MHD instabilities.[30]

3.5. *High redshift afterglows*

GRB 050904, at the unprecedented redshift of $z = 6.29$,[10] had an X-ray brightness exceeding for a day that of the brightest X-ray quasars,[11] and its initial O/IR brightness was comparable to that of the record-holding ($m_V \sim 9$) optical flash in GRB 990123.[8,12] Besides a few bursts with extremely bright prompt optical emission, there have been a score of other early optical flashes with more modest initial brightnesses $m_v \gtrsim 14$, discussed in §3.1. The unprecedented brightness in O/IR and in X-rays, and the high redshift of GRB 050904 underline the potential of GRB for investigations of the IGM, the star formation rate and early galactic environments,[95–97] even out to redshifts $\sim 20 - 30$, if present there. Detailed multi-wavelength light curve and spectral fits of GRB 050904[98] allow to probe not only the afterglow mechanism but also the external environment at redshifts close to that of reionization. Observations suggest that the metallicity of long GRB host galaxies is lower than in average massive star forming galaxies.[99–101] This has implications for the redshift distribution of GRB.[102,103] A low progenitor metallicity could promote fast rotation of the core,[28] a prerequisite for the collapsar model of long bursts.

The use of GRB for cosmology tests may be possible, using empirical correlations between burst properties. E.g. between the photon spectral peak energy E_{pk} and the apparent isotropic energy E_{iso} one has[104] $E_{pk} \propto E_{iso}^{1/2}$. The dispersion of this correlation is substantial, so its usefulness is limited (see also.[75] A tighter correlation exists between E_{pk} and the collimation-corrected total energy E_j, which relies on (scarce) optical afterglow light-curve breaks to get the collimation.[70,105–107,109] This has has been advocated as a cosmological tool[108,109] [c.f.[110]]. One problem is that this mixes a prompt emission quantity (the gamma-ray E_{pk}) with an optical afterglow quantity determined a day or so later. Selection effects, and the dependence of the results on model assumptions are also a problem. The latter may be circumvented by relying only on observables, such as E_{peak}, the fluence (or peak flux) and the break time t_{br}.[36,111] The most promising correlation so far is one that involves purely prompt (gamma-ray) quantities, between the rest-frame E_{pk}, L_{iso} and $t_{0.45}$, namely $L_{iso} \propto E_{pk}^{1.6} t_{0.45}^{-0.5}$.[112] The $t_{0.45}$ measures the variability of the gamma-ray light curve, which roughly scales with the duration. This was used to extend the Hubble diagram, combining 199 SNe Ia and 19 GRBs, up to $z \lesssim 5$, in a Bayesian analysis to circumvent the circularity implicit in the use of the cosmological models tested.[113]

3.6. *GRB-SN*

The first spectroscopic evidence for a long GRB-supernova association was in GRB980425/SN1998bw.[114,115] This was a peculiar type Ib/c supernova, where the associated GRB properties seemed the usual, except that the extremely small redshift $z \sim 0.0085$ implies an extremely low isotropic equivalent energy $E_\gamma \sim 10^{48}$ erg. Using SN1998bw as a template, other possible associations were reported through photometric detection of reddened bumps in the optical afterglow. A second, completely unambiguous GRB-SN association was that of GRB 030329/SN2003dh, at a redshift $z = 0.169$,[116,117] again a SN Ib/c. The delay between GRB 030329 and SN 2003dh is less than two days, compatible with both events being simultaneous.[117] Other pre-Swift GRB-SN associations are discussed in.[118,119]

More recently, Swift observed with all three instruments, BAT, XRT and UVOT, an unusually long ($\sim$ 2000 s), soft burst, GRB 060218,[49] which was associated with SN2006aj, a very nearby ($z = 0.033$) type Ic supernova.[50,51,120–123] The supernova light curve peaked earlier than most known supernovae, and its time origin can also be constrained to be within less than a day from the GRB trigger. This is the first time that a connected GRB and supernova event has been observed starting in the first $\sim$ 100 s in X-rays and UV/Optical light, and the results are of great interest. The early X-ray light curve shows a slow rise and plateau followed by a drop after $\sim 10^3$ s, with a power law spectrum and an increasing black-body like component which dominates at the end. The most interesting interpretation involves shock break-out of a semi-relativistic component in a WR progenitor wind[49] (c.f.[124,125]). After this a more conventional X-ray power law decay follows, and a UV component peak at a later time can be interpreted as due to the slower supernova envelope shock. Another GRB/SN detection based on a Swift detection is GRB 050525A.[2]

3.7. *Short bursts*

The X-ray afterglows of short bursts, first detected by Swift, resemble those of long bursts, as expected if both long and short burst afterglows are described by the fireball shock paradigm. The X-ray light curve temporal slope is, on average, that expected from the forward shock, and in two short bursts there is evidence for a light curve break which could be due to jet effects.[31,33] A steep initial decay, flattening and flares in X-rays (e.g. in GRB 050724) are also similar to Swift long burst features. However, while similar to zeroth order, the first order differences are interesting. The average isotropic energy is $\sim$ 100 times smaller, and the jet opening angle (based on two breaks) is $\sim$ 2 times larger[31] than in long bursts. This is natural if short bursts arise from compact mergers, since NS-NS and NS-BH mergers are expected to lead to lower total jet energies, and broader jets (due to the lack of a collimating stellar envelope). The identified host galaxies (half ellipticals, half irregulars with low SFR) also conform to the notion that they arise in old populations compatible with (but not necessarily implying) compact mergers.

Two challenges posed by the Swift short burst afterglows are, in some cases, a long, soft tail of the prompt emission (although these would have been unequivocally classified as short by BATSE); and the strength and late occurrence of X-ray flares. The extended prompt soft tails ($\lesssim 100s$) may be possible in black hole - neutron star mergers,[32] for which analytical and numerical arguments suggest ([5,35]) a more complex and extended accretion history than for NS-NS mergers. The simplest interpretation for the flares may be refreshed shocks, if the rise and decay times are moderate. However, in the GRB 050724 flare at 10^4 s, the energy in the slow late material would need to be ten times larger than in the prompt emission. A possible mechanism might be temporary choking up of an MHD outflow[30] (c.f.[34]). Such MHD effects could plausibly also explain the prompt soft tails. However, significant flares at $t \gtrsim 10^5$ s remain a challenge.

3.8. *Long-short classification*

The soft prompt emission tail in bursts which BATSE would have classified as short is more glaring in GRB 060614,[1] where its energy content is five times that in the prompt hard initial spike. In retrospect, long soft tails are present also in a fraction of BATSE bursts,[4] although usually not as bright as in this case. The candidate host of GRB 060614 is a dwarf galaxy at $z = 0.125$, and any associated SN, if present, would have a luminosity upper limit $\lesssim 100$ smaller than any previous SN.[2,3] Similar negative results were obtained in SN searches on other Swift short bursts, as expected from a compact merger origin. Also,h Swift observations indicate that short bursts arise in old populations. Interestingly, the time lags (relative delays between hard and soft components of the same pulses) for this, and other Swift short bursts, are zero whithin the errors,[1] as found for short bursts in general.[4] This reinforces the view that it belongs to the generic class of short bursts. The only spoiler could be the energetic soft long tail, which is suggestive of a long burst or XRF. This has been investigated by,[135] who show that, if short bursts satisfy the Amati relation (as the appear to do), luminous bursts such as GRB 060614 would appear, when scaled down to lower luminosities, completely similar to other canonical short bursts. The nature of the apparent overlap between the two traditional populations is an issue which clearly requires further exploration.

I am grateful to X-Y. Wang. D. Fox and L-J. Gou for collaborations, and NASA NAG5-13286 for support.

References

1. Gehrels, N, et al, 2006, Nature in press (astro-ph/0610635)
2. DellaValle, M, et al, Nature, 2006, in press (astro-ph/0608322)
3. Fynbo, J, et al, 2006, Nature in press (astro-ph/0608313)
4. Norris, J.P and Bonnell, J.T, 2006, ApJ 643:266
5. Faber, J et al, 2006, ApJ, 641:L93
6. Burrows, D, 2006, AIPC 836:215
7. Roming, P & Mason, K., 2006, AIPC 836:224
8. Haislip, J, et al, 2006, Nature, 440:181
9. Berger, E, et al, 2005, ApJ 634:501
10. Kawai, N. et al, 2005, astro-ph/0512052
11. Watson, D. et al, 2006, ApJ 637:L69
12. Boër, M, et al, 2006ApJ 638:L71
13. Frail, D et al., ApJ 562:L155
14. Berger, E, Kulkarni, SR and Frail, D, 2003, ApJ 590:379
15. Kumar, P and Panaitescu, A, 2000, ApJ 541:L51
16. Zhang, B. et al, 2006, ApJ 642:354
17. Nousek, J, et al, 2006, ApJ, 642:389
18. Panaitescu, A, et al, 2006, MNRAS 366:1357
19. Kobayashi, S, et al, 2006, ApJ in press (astro-ph/0506157)
20. Rees, MJ and Mészáros, P, 1998, ApJ, 496:L1
21. Sari, R and Mészáros, P, 2000, ApJ, 535:L33
22. O'Brien, P. et al, Astron.Ap. in press (astro-ph/0601125)
23. Zhang, B and Mészáros, P, 2001, ApJ 552:L35
24. Granot, J and Kumar, P, 2006, MNRAS 366:L13
25. Burrows, D, et al, 2005a, Science, 309, 1833
26. Ioka, K, et al, 2006, Astron.Ap. in press (astro-ph/0511749)
27. Dermer, C, 2006, AIPC 836:97
28. Woosley, S & Heger, A, 2006, AIPC 836:398
29. Perna, R, Armitage, and Zhang, B, 2006, ApJ 636:L29
30. Proga, D and Begelman, M, 2003, ApJ, 592:767
31. Panaitescu, A, 2006, MNRAS 367:L42
32. Barthelmy, S., 2005, Nature, 438:994
33. Berger, E, et al, 2005, ApJ, 634:501
34. Van Putten, M and Ostriker, A, 2001, ApJ 552:L31
35. Davies, M, Levan, A, King, A, 2005, MNRAS 365:54
36. Liang, EW and Zhang, B, 2006, ApJ 638:L67
37. Mészáros, P and Rees, MJ, 1997, ApJ 476:232
38. B. Zhang, S. Kobayashi and P. Mészáros, 2003, ApJ, 589:861
39. Ryde, F, 2004, ApJ, 614, 827
40. Piran, T. 2005, Rev. Mod. Phys. 76, 1143
41. Mészáros, P, 2006, Rep.Prog.Phys. 69:2259
42. Band, D, 2006, ApJ 644:378
43. Gehrels, N. et al. 2004, ApJ 611:1005
44. Burrows, D. N. et al., 2005, Space Sci. Rev. 120:165
45. Roming, P. et al., 2005, Procs. SPIE 5165:262

46. Jakobsson, P. et al., 2006, astro-ph/0602071
47. Bagoly, Z. et al., 2006, Astron.Ap., 453:797
48. Roming, P. et al. 2006a, ApJ in press (astro-ph/0509273)
49. Campana, S. et al., 2006, Nature, 442:1008
50. Modjaz, M. et al. 2006, ApJ in press (astro-ph/0603377)
51. Mirabal, N., et al, 2006, ApJ 643:L99
52. Soderberg, A. M. et al., 2006, astro-ph/0604389
53. Della Valle, M. et al., 2006a, ApJ 642:L103
54. Soderberg, A. M. et al., 2006, astro-ph/0601455
55. Band, D. et al., 1993, ApJ 413:281
56. Lyutikov, M and Blandford, R, 2003, astro-ph/0312347
57. Preece, R. D., et al, 2000, ApJ(Supp) 126:19
58. Granot, J., Piran, T., and Sari, R., 2000, ApJ 534:L163
59. Panaitescu, A. and Mészáros, P., 2000, ApJ 544:L17
60. Medvedev A. 2000. ApJ 540:704
61. Medvedev A, 2006, ApJ 637:869
62. Lloyd N & Petrosian V. 2001a. ApJ 543:722
63. Lloyd-Ronning, N & Petrosian, V, 2002, ApJ 543:722
64. MészárosP & Rees MJ. 2000b, ApJ 530:292
65. Pe'er, A., & Waxman, E. 2004, ApJ, 613:448
66. Pe'er, A., & Waxman, E. 2005, ApJ, 628:857
67. Rees, MJ and Meszaros, P, 2005, ApJ, 628:847
68. Pe'er, A., Mészáros, P., and Rees, M. J., 2005, ApJ 635:476
69. Pe'er, A., Mészáros, P., and Rees, M. J., 2006, ApJ 642:995
70. Ghirlanda, G; Ghisellini, G; Lazzati, D, 2004 ApJ 616:331
71. Thompson, C, Mészáros, P & Rees, MJ, 2006, astro-ph/0608282
72. Akerlof, C. et al., 2000, Nature 398:400
73. Sari, R & Piran, T, 1999, ApJ 517, L109
74. Mészáros, P and Rees, MJ, 1999, MNRAS, 306:L39
75. Nakar, E. and Piran, T., 2005, ApJ 619:L147
76. Vestrand, W. T. et al., 2005, Nature 435:178
77. Blake, C. H. et al., 2005, Nature 435:181
78. Vestrand, W. T. et al., 2006, astro-ph/0605472
79. Rykoff, E. S. et al., 2005, ApJ 640:L121
80. Quimby, R. M. et al., 2006, ApJ 640:402
81. Yost, S. A. et al., 2006, ApJ 636:959
82. Rykoff, E. S. et al., 2004, ApJ 601:1013
83. Rykoff, E. S. et al., 2006, ApJ 638:L5
84. Kobayashi, S., 2000, ApJ 545:807
85. Mészáros, P, Ramirez-Ruiz, E, Rees, MJ and Zhang, B, 2002, ApJ 578:812
86. Nakar, E. and Piran, T., 2004, MNRAS 353:647
87. McMahon, E., Kumar, P., and Piran, T., 2006, MNRAS 366:575
88. Mészáros, P, Rees, MJ & Papathanassiou, H, 1994, ApJ, 432:181
89. Blustin, A. J. et al., 2006, ApJ 637:901
90. Klotz, A., et al, 2006, Astron.Ap 451:L39
91. Jelínek, M. et al., 2006, astro-ph/0606004

92. Fan, Y. Z., Zhang, B., and Wei, D. M., 2005, ApJ 628:L25
93. Wei, D. M., Yan, T., and Fan, Y. Z., 2006, ApJ 636:L69
94. Wei, D. M., 2006, astro-ph/0605016
95. Lamb, DQ & Reichart, D, 2000, ApJ 536, 1L
96. Ciardi B and Loeb A, 2000, ApJ 540:687
97. Gou, LJ, Mészáros, P, Abel, T and Zhang, B, 2003, ApJ, 604:508
98. Gou, LJ, Fox, D and Mészáros, P, 2006, ApJ, subm (astro-ph/0612256)
99. Le Floc'h, et al, 2003, Astron.Ap. 400:499
100. Le Floc'h, et al, AIPC 836:528
101. Stanek, K, 2006, ApJ subm (astro-ph/0604113)
102. Natarajan, P, et al, 2005, MNRAS 364:L8
103. Fruchter, A, et al, 2006, Nature 441:463
104. Amati, L et al, 2002, AA, 390:81
105. Ghirlanda, G; et al, 2004, ApJ 613:L13
106. Dai, ZG, Liang, EW, Xu, D, 2004, ApJ 612:L101
107. Xu, D, Dai, ZG, Liang, EW, 2005, ApJ, 633:603
108. Lamb, D.Q., et al, 2005b, astro-ph/0507362
109. Ghirlanda, G; et al, 2005, AIPC 801:123
110. Friedman, Andrew S.; Bloom, Joshua S., 2005, ApJ 627:1
111. Nava, L, et al, 2006, Astron.Ap. in press (astro-ph/0511499)
112. Firmani, C, et al, 2006a, MNRAS 370:185
113. Firmani, C, et al., 2006b, astro-ph/0605430
114. Galama, T. et al., 1998, Nature, 395:670
115. Kulkarni, S et al., 1998b, Nature, 395:663
116. Stanek, K et al., 2003, ApJ 591:L17
117. Hjorth, J et al., 2003, Nature 423:847
118. Zeh, A, Klose, S, Hartmann, D, 2004, Astron.Ap., 422:113
119. Zeh, A, Klose, S, Kann, D, 2006, ApJ, 637:889
120. Masetti, N, Palazzi, E, Pian, E, Patat, F, 2006, 6GCN 4803
121. Pian, E, et al, 2006, Nature, subm (astro-ph/0603530)
122. Sollerman, J, et al, 2006, AA sub (astro-ph/0603495)
123. Cobb, B, et al, 2006, ApJL sub (astro-ph/0603832)
124. Fan, Y, Piran, T, Xu, D, 2006, astro-ph/0604016
125. Tan, JC, Matzner, CD, McKee, CF, 2001, ApJ, 551:946
126. Barthelmy, S. D. et al., 2005, Space Sci. Rev. 120:143
127. Kobayashi, S and Zhang, B, 2006, astro-ph/0608132
128. Woosley, S., 1993, Ap.J., 405, 273
129. Paczyński , B., 1998, ApJ, 494:L45
130. Fryer C, Woosley S & Hartmann D, 1999. ApJ 526:152
131. Paczyński , B., 1986, ApJ, 308:L43
132. Goodman, J., 1986, ApJ, 1986, 308, L47
133. Eichler D, Livio M, Piran T & Schramm D.1989. Nature 340:126
134. Mészáros, P & Rees, MJ, 1997b, ApJ, 482, L29
135. Zhang, B, et al, ApJ (Lett.) in press (astro-ph/0612238)

MODELING THE MULTIWAVELENGTH SPECTRA AND VARIABILITY OF 3C 66A IN 2003–2004

M. Joshi* and M. Böttcher

Astrophysical Institute, Department of Physics and Astronomy, Ohio University, Athens, Ohio 45701, USA
**E-mail: joshi@helios.phy.ohiou.edu*
www.ohiou.edu

The BL Lac object 3C 66A was the target of an intensive multiwavelength monitoring campaign organized in 2003–2004. During the campaign, its spectral energy distribution (SED) was measured and flux measurements from radio to X-ray frequencies as well as upper limits in the very high energy (VHE) γ-ray regime were obtained. Here, we reproduce the SED and optical spectral variability pattern observed during our multiwavelength campaign using a time-dependent leptonic jet model. Our model could successfully simulate the observed SED and optical light curves and predict an intrinsic cutoff value for the VHE γ-ray emission at $\sim$ 4 GeV.

Keywords: galaxies; active; BL Lacertae objects; 3C 66A.

1. Introduction

Blazars constitute the most extreme class of Active Galactic Nuclei (AGN) and exhibit the most violent non-transient high-energy phenomena observed so far. They are primarily characterized on the basis of their non-thermal continuum spectra and radio jets with individual components often exhibiting apparent superluminal motion. The broadband spectra of blazars consists of two broad spectral components that are associated with non-thermal emission processes. The synchrotron emission from non-thermal electrons in a relativistic jet produces the low-energy component whereas the high-energy component is attributed either to the Compton upscattering of low energy radiation by the synchrotron emitting electrons (for a recent review see, e.g., 1) or the hadronic processes initiated by relativistic protons co-accelerated with the electrons [6,7]. Blazars are often known to exhibit variability at all wavelengths, varying on time scales from months, to a few days, to even less than an hour in some cases.

The blazar 3C 66A is a low-frequency peaked (or radio selected) BL Lac object (LBL) with a relatively uncertain redshift determination of $z = 0.444$ [4]. It has previously exhibited rapid microvariability at optical and near infrared and has been suggested as a promising candidate for detection by H.E.S.S., MAGIC, or VERITAS [5]. In this paper, we use a leptonic jet model to reproduce the broadband SED of 3C 66A and the observed optical variability pattern and make predictions regarding the intrinsic cutoff value of the spectrum at VHE γ-rays.

2. Model description and model parameters

A one-zone homogeneous leptonic jet model was used to simulate the SED as well as the observed optical variability pattern of 3C 66A. According to the model, a population of ultrarelativistic non-thermal particles (electrons and positrons) is continuously injected into a spherical emitting volume (the "blob") of comoving radius R_b at a time-dependent rate. The injected population follows a single power law distribution described by a particle spectral index p, comoving density n_e and low- and high-energy cutoffs γ_1 and γ_2, respectively, such that $n_e(\gamma) = n_0\gamma^{-p}$ for $\gamma_1 \leq \gamma \leq \gamma_2$. The blob carries a randomly oriented magnetic field B of uniform strengt , which is determined by an equipartition parameter $e_B \equiv u_B \ / \ u_e$ (in the comoving frame), where u_B is the magnetic field energy density and u_e is the electron energy density. At a height z_0, above the plane of the disk, the electron population is injected initially and the emitting region starts to travel relativistically with a speed $v/c = \beta_\Gamma = (1 - 1/\Gamma^2)^{1/2}$ along the jet. The jet is directed at an angle θ_{obs} with respect to the line of sight and the Doppler boosting of the emission region with respect to the observer's frame is determined by the Doppler factor $\delta = [\Gamma(1 - \beta_\Gamma \cos\theta_{obs})]^{-1}$. As the emission region travels outward along the jet, the electron population in the blob loses energy via synchrotron emission, Compton upscattering of synchrotron photons (SSC) and/or Compton upscattering of external photons (EIC). The evolution of electron and photon population inside the emission region is governed by equations (4) and (5) of 3.

The model independent parameters given in equation (4) of 2 were used to form a base set of input parameters to reproduce the quiescent and the flaring state of 3C 66A. Approximately, 350 simulations were carried out to study the effects of variations of various parameters, such as γ_1, γ_2, p, B and Γ, on the resulting broadband spectra and light curves. The various model parameters used to simulate the two states of 3C 66A, using a pure SSC emission process, are listed in Table 1. A Doppler factor of $\delta = \Gamma = 24$ and a viewing angle of $\theta_{\rm obs} = 2.4^{\rm o}$ resulted in a satisfactory fit to the quiescent state of 3C 66A. [2]. The quiescent state was simulated such that it did

Table 1. Model Parameters used to reproduce the quiescent and flaring state of 3C 66A as shown in Figure 1.

Fit	L_{inj} [10^{41} ergs/s]	γ_1 [10^3]	γ_2 [10^4]	p	Profile	e_B	B [G]	Γ	R_b [10^{15} cm]	θ_{obs} [deg]
1	2.7	1.8	3.0	3.1	——	1	2.4	24	3.59	2.4
2	8.0	2.1	4.5	2.4	Gaussian	1	2.8	24	3.59	2.4

Note: L_{inj}: luminosity of the injected electron population in the blob, $\gamma_{1,2}$: low- and high-energy cutoffs of electron injection spectrum, p: particle spectral index, Profile: flare profile used to simulate the optical variability pattern, e_B: equipartition parameter, B: equipartition value of the magnetic field, Γ: bulk Lorentz factor, R_b: comoving radius and θ_{obs}: viewing angle.

not overpredict the X-ray photon flux as X-ray photons are expected to be dominated by the flaring episodes. On the other hand, the flaring state was reproduced such that the simulated time-averaged spectrum passes through the observed time-averaged optical and X-ray data points. This was achieved by varying γ_1, γ_2 and p. The effect of EIC mechanism on the high-energy component of the spectra of 3C 66A has not been considered yet and is a work in progress.

3. Results and discussion

Figure 1 shows the simulated SED of 3C 66A, for the quiescent and the flaring state. The quiescent state is a simulation of the state observed around 1st October 2003 whereas the flaring state is the reproduction of a generic 10 day flaring period corresponding to the timescale of several major outbursts that were observed during the campaign. The flaring state was reproduced by varying individual input parameters between the values for quiescent and flaring states with a profile that was Gaussian in time. The change in the value of p, in our simulations, from 3.1 to 2.4 might indicate a possible change in the B-field orientation or an interplay between the 1st and 2nd order Fermi acceleration that is making the particle spectra harder.

Figure 2 shows the simulated time-averaged spectrum of 3C 66A in the flaring state. As can be seen, the high energy end of the synchrotron component passes through the time averaged X-ray data. This shows that the soft X-ray photons are produced from synchrotron emission during flaring

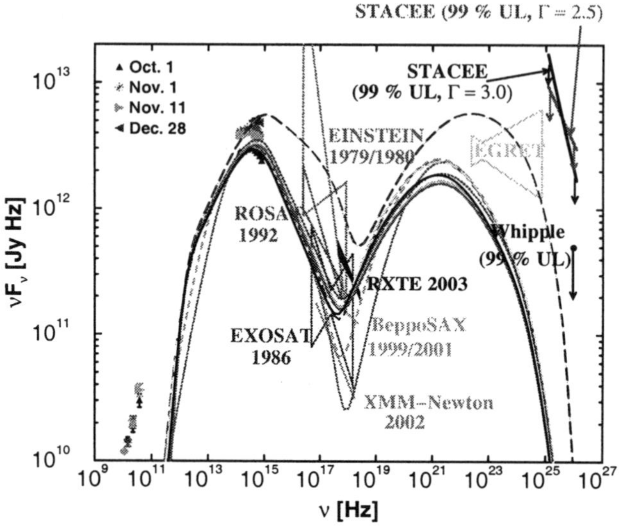

Fig. 1. Simulation of the quiescent state of 3C 66A observed around October 1st 2003 and the flaring state for a generic 10 day flare corresponding to the timescale of several major outbursts observed in the optical regime during the campaign. The black solid line indicates the instantaneous spectrum generated after the system attains equilibrium in the quiescent state. The low-energy component of the quiescent state peaks in the optical at $\nu_{\rm syn} \approx 4.8 \times 10^{14}$ Hz whereas the high-energy SSC component peaks in the MeV regime at $\nu_{\rm SSC} \approx 1.6 \times 10^{21}$ Hz. The synchrotron cooling timescale in the observer's frame is ≈ 1.2 hours, which is on the order of the observed minimum optical variability timescale of 2 hours. The rest of the curves show the instantaneous spectra in the flaring state at several different times in the observer's frame, for e.g., long-dashed black line (~8th day, highest state attained by the system during flaring) and red solid line (~22nd day, equilibrium state reached by the system after the flaring episode is over). The synchrotron component of the flaring state peaks at $\nu_{\rm syn} \approx 1.1 \times 10^{15}$ Hz and the SSC component peaks at $\nu_{\rm SSC} \approx 2.7 \times 10^{22}$ Hz. The SSC component of this state cuts off at $\nu_{\rm SSC,cutoff} \approx 2.3 \times 10^{24}$ Hz and the synchrotron cooling timescale is ≈ 37 minutes.

whereas the harder X-ray photons come from the SSC mechanism with the expected spectral hardening taking place at ~ 7 keV. The high energy component, due to SSC emission, for the time-averaged spectrum cuts off at ~ 4 GeV and our modeling results predict that the object is within the sensitivity limits of MAGIC, VERITAS and GLAST.

Figure 3 is a hardness intensity graph that indicates that the object follows a positive correlation of becoming harder in B-R while getting brighter in the R band, which agrees well with the observed optical variability pattern. The simulated variability amplitude in the R band (0.55 mag) also

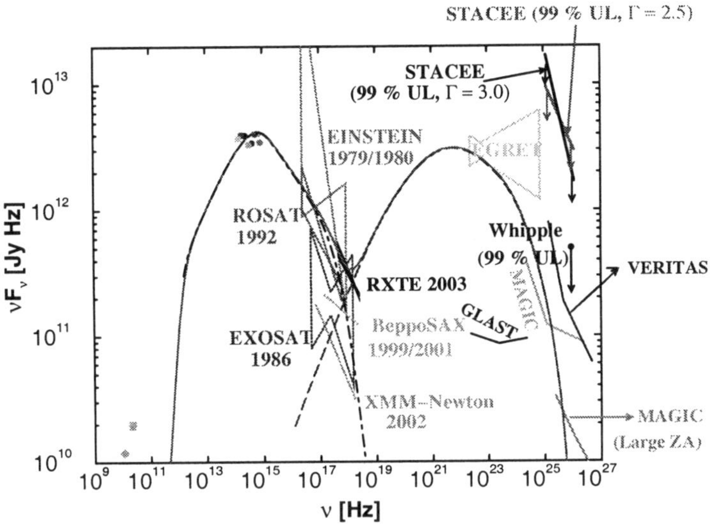

Fig. 2. Time-averaged spectral energy distribution of 3C 66A around a flare as shown in Figure 1. The filled colored circles are the time-averaged optical and IR data points for the entire campaign period and the "RXTE 2003" denotes the time-averaged X-ray data points. The dot-dashed black line is the contribution from the synchrotron component only whereas the long-dashed blue line indicates the contribution of the SSC component only. The time-averaged synchrotron component peaks at $\nu_{\rm syn} \approx 7.2 \times 10^{14}$ Hz whereas the time-averaged SSC component peaks at $\nu_{\rm SSC} \approx 5.3 \times 10^{21}$ Hz. The green, maroon and magenta lines indicate the sensitivity limits for an observation time of 50 hours for MAGIC, VERITAS and MAGIC (Large Zenith Angle) respectively whereas, the black line indicates the sensitivity limit for GLAST for an observation time of 1 month.

matches the observed value (0.3 - 0.5 mag) for a 10 day period outburst.

4. Summary

A detailed analysis of the data of 3C 66A was carried out using a one-zone time-dependent homogeneous leptonic jet model. The simulations yielded a satisfactory fit to the observed SED in the quiescent as well as the flaring state and could successfully reproduce the observed optical variability pattern. According to the simulations, the production of hard X-ray and VHE photons is dominated by the SSC mechanism throughout whereas the soft X-ray photons start out with the dominance of the SSC mechanism during the quiescent state and later on get taken over by the synchrotron mechanism during the flaring state. The synchrotron component is expected to cut off near 7 keV whereas the SSC component cuts off at ~4 GeV yielding

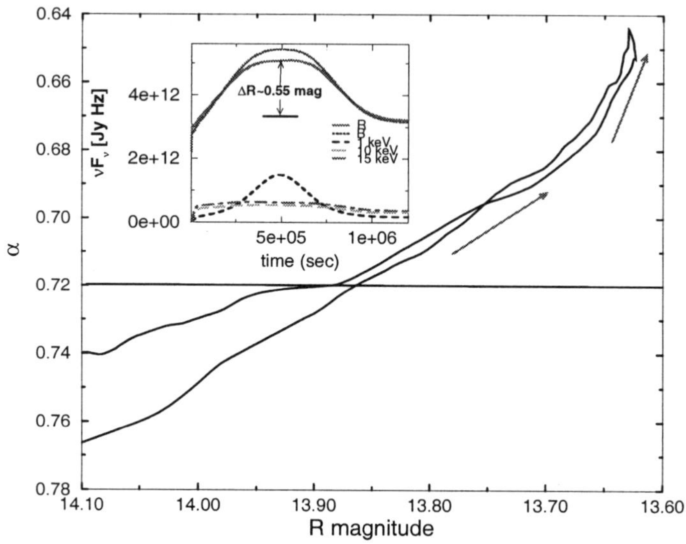

Fig. 3. The simulated hardness-intensity diagram indicates a positive correlation between the R- and B-band for an outburst lasting for $\sim$ 10 days. The object becomes brighter in B and harder in B-R as shown by the red arrows. The upturn takes place at B-R $\approx$ 0.72 mag where the flux in B equals that in R (corresponding to $\alpha_{\rm BR} = 0$). The inset figure shows the simulated light curves for various energy bands. The simulated variability in the R band is $\approx$ 0.55 mag as indicated by the black arrows.

an intrinsic cutoff value at VHE for this object. This puts the object well within the observational range of MAGIC, VERITAS and GLAST. In the ongoing work, the possible presence of an external inverse Compton component is being evaluated (Joshi & Böttcher, in preparation), which may substantially enhance the level of expected γ-ray emission.

References

1. Böttcher, M., 2006, in proc. "The Multi-Messenger Approach to High-Energy Gamma-Ray Sources", Barcelona, Spain, 2006, Astroph. & Space Sci., in press
2. Böttcher, M., et al., 2005, ApJ, 631, 169
3. Böttcher, M., & Chiang, J., 2002, ApJ, 581, 127
4. Bramel, D. A., et al., 2005, ApJ, 629, 108
5. Costamante, L., & Ghisellini, G., 2002, A&A, 384, 56
6. Mücke, A., & Protheroe, R. J., 2001, Astropart. Phys., 15, 121
7. Mücke, A., et al., 2003, Astropart. Phys., 18, 593

HIGH ENERGY SIGNATURES OF THE POST-ADIABATIC SUPERNOVA REMNANTS

I. O. Telezhinsky* and B. I. Hnatyk**

Astronomical Observatory of Kiev University,
Observatorna str. 3, Kiev 04053, Ukraine
** E-mail: igor_t@online.com.ua*
*** E-mail: hnatyk@observ.univ.kiev.ua*

Between the well known adiabatic and radiative stages of the Supernova remnant (SNR) evolution there is, in fact, a transition stage with a duration comparable to the duration of adiabatic one. Physical existence of the transition stage is motivated by cooling of some part of the downstream hot gas with formation of a thin cold shell that is joined to a shell of swept up interstellar medium (ISM). We give an approximate analytical method for full hydrodynamic description of the transition stage. On its base we investigate the evolution of X-ray and γ-ray radiation during this stage. The role of the transition stage in cosmic ray (CR) re-acceleration is discussed as well.

Keywords: Supernova Remnants: Evolution, X-ray radiation, γ-ray radiation, cosmic ray acceleration. Interstellar medium: general

1. Introduction

During their lifespan SNRs are thought to go through three main stages of evolution: free expansion, adiabatic and radiative (see Ref. 7 and Refs. therein).Transition between phases is accompanied by change of basic hydrodynamic characteristics of plasma flow in an SNR. Free expansion phase ends when the stellar ejecta slows down and transforms its energy into the strong nonradiative (adiabatic) shockwave in the ISM. With time, radiative losses of shocked plasma become important and dominate close to the shock front, where the plasma density is at maximum. As a consequence, a thin relatively cold dense shell of newly shocked, heated and quickly cooled ISM is formed, manifesting the beginning of classical radiative stage, when the hot internal plasma pushes the cold shell of swept up ISM (so called "pressure-driven snowplow" (PDS) model[8]).

However, numerical calculations[1,2] show that PDS approximation is not appropriate starting from the time of cooling of the first gas element and formation of the infinitesimal shell. It is because during some period of time besides swept up ISM a considerable portion of the internal hot gas is joining the shell. And only when the process of fast cooling and joining to the cold shell of the internal hot plasma ceases, PDS approximation is appropriate. We call this period of SNR evolution "transition stage" and propose its analytical description. We also show that conversion of the hot internal plasma to the cold shell results in evolution of X-ray and γ-ray radiation of SNRs during transition stage.

2. Hydrodynamic model of the transition stage

Here we generalize our method[6] for analytical description of adiabatic stage of SNR evolution for description of transition stage. The method is based on simulteneous usage of Lagrangian and Euler coordiantes for the plasma flow description. After the end of transition stage we use PDS approximation for description of radiative stage and thus we are able to describe the whole SNR evolution.

2.1. *Origin and dynamics of the thing shell during transition phase*

Numerical simulations[1,2] show that deviation from self-similarity starts at the time t_{tr} when the cooling time of the gas $t_c = E(T_s, \rho_s)/\Lambda(T_s, \rho_s)$ (where $\Lambda(T_s, \rho_s)$ and $E(T_s, \rho_s)$ are the emissivity and the thermal energy of the plasma just behind the shock) is comparable with the age of the SNR:

$$t_{tr} = t_c = 2.9 \times 10^4 E_{SN,51}^{4/17} n_H^{-9/17}\, yr, \tag{1}$$

where $E_{SN,51}$ is the explosion energy in 10^{51} ergs, n_H and n_e are the ISM hydrogen and electron number density.

Radiative losses lead to rapid formation the of cold dense shell near the front. The shell increases because the hot SNR gas cools in the reverse shock when it rushes at and joins the inner boundary of the shell and because the shell sweeps up the ISM. Transition phase ends when the hot gas stops cooling effectively and no more replenishes the shell. Because of nonstationarity of the process and complexity of conditions at the reverse shock front, the duration of transition stage cannot be deduced analytically. We use the numerical results[1] saying that cooling is important for the hot plasma within outer five percent ($\alpha = 0.05$) of the shockwave radius at the beginning of transition phase $\Delta r = \alpha R_{tr}$. Parameter α is the only free parameter of our model.

So, we model complicated processes during transition stage as follows. We take the time t_{tr} for the beginning of transition stage, when the first cold gas element of the shell appears at the front of the adiabatic shockwave with the temperature $T_{sh} = T_{ISM} = 10^4 K$, pressure P_{sh}, density ρ_{sh} and velocity V_{sh}. From the balance of external and internal pressure on the shell $P_{sh} = P_{dyn}$ or $\rho_{ISM} V_{sh}^2 = \rho_{sw}(v_{sw} - V_{sh})^2$, we derive the velocity of the shell:

$$V_{sh} = \frac{1}{2} D_{sw}(t_{tr}) = const \tag{2}$$

for adiabatic index $\gamma = 5/3$, where $\rho_{ISM} = \mu n_H m_H$ is the ISM density, μ is the molar mass, ρ_{sw} and v_{sw} are the plasma density and velocity at the shock front for $t = t_{tr}$. Dynamic pressures are changing slightly with time, so we take that the shell velocity V_{sh} is constant during transition phase and is determined by Eq. (2). We also assume that during transition stage the small pressure gradient inside the SNR results in conserving the velocity of each plasma element, unless and until it joints the shell:

$$v(a,t) = \begin{cases} v(a, t_{tr}) & \text{if } 0 < a < a_c(t) \\ V_{sh} & \text{if } a_c(t) < a < R_{tr} \end{cases} \tag{3}$$

where Lagrangian coordinate $a_c(t)$ of the gas element is determined from condition that the gas element reaches the cold shell and cools at the time t: $R_{tr} - r(a_c, t_{tr}) = (v(a_c, t_{tr}) - V_{sh})(t - t_{tr})$. For the end of transition phase $a_c(t_{sf}) = a_{min}$ and is determined from condition $r(a_{min}, t_{tr}) = r_{min} = (1 - \alpha) R_{tr}$. Using it for cooling of the outermost gas element with Euler coordinate r_{min} we can derive the duration of transition phase in our model:

$$\Delta t = t_{sf} - t_{tr} = \frac{\alpha R_{tr}}{v(r_{min}, t_{tr}) - V_{sh}} \tag{4}$$

From Eq. (4) we can see that it is comparable to the SNR age.

2.2. *Hot gas parameters inside the shell*

For the time $t_{tr} < t < t_{sf}$ the velocity of the gas element with $0 < a < a_c(t)$ is given by Eq. (3), so for the Euler coordinate $r(a,t)$ we have:

$$r(a,t) = \begin{cases} r(a, t_{tr}) + v(a, t_{tr})(t - t_{tr}) & \text{if } 0 < a < a_c(t) \\ R_{sh} & \text{if } a_c(t) < a < R_{tr} \end{cases} \tag{5}$$

The density distribution $\rho(a,t)$ we find from continuity condition:

$$\rho(a,t) = \rho(a, t_{tr}) \left(\frac{r(a, t_{tr})}{r(a,t)} \right)^2 \frac{dr(a, t_{tr})}{dr(a,t)} \tag{6}$$

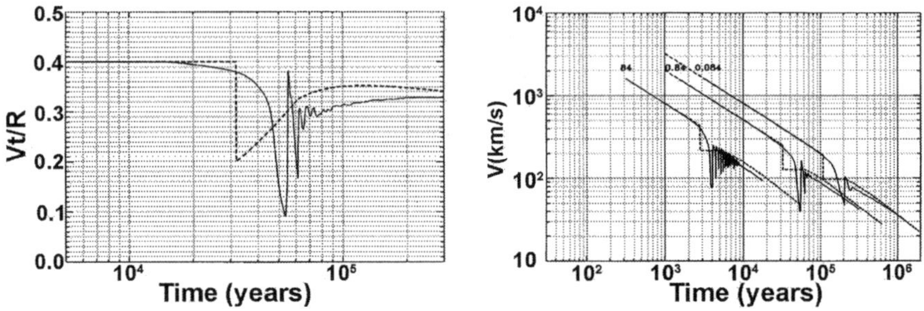

Fig. 1. Comparison of the proposed method (dash) with numerical simulation[1] (solid) for the explosion energy $E_{SN} = 10^{51}$ ergs. Left - evolution of deceleration parameter $m = Vt/R$ for the ISM number density $n_H = 0.84\ cm^{-3}$. Right - the SNR front velocity evolution during adiabatic, transition and radiative stages for different ISM densities.

the hot gas pressure is now:

$$P(a,t) = P(a,t_{tr})\left(\frac{\rho(a,t)}{\rho(a,t_{tr})}\right)^{\gamma} \tag{7}$$

and the temperature:

$$T(a,t) = \frac{\mu P(a,t)}{R_g \rho(a,t)}, \tag{8}$$

where μ is the molar mass and R_g is the absolute gas constant. Thus we give the full description of the hot gas inside the shell.

2.3. *Cold shell gas parameters*

Starting from the time t_{tr} when the first cold element appeared at the distance R_{tr} the mass of the shell increases because the cooled hot gas joins the shell and the shell sweeps up the ISM:

$$M_{sh,in}(t) = 4\pi \int_{a_c(t)}^{R_{tr}} \rho_{ISM} a^2 da \tag{9}$$

$$M_{sh,out}(t) = 4\pi \int_{R_{tr}}^{R_{sh}} \rho_{ISM} a^2 da \tag{10}$$

The temperature of the cold shell gas equals the ISM temperature: $T_{sh}(t) = T_{ism} = 10^4 K$ and its pressure is $P_{sh} = P_{dyn} = \rho_{ISM} V_{sh}^2$. From equation of state we have: $\rho_{sh} = \rho_{ISM} M_{iso}^2$, where M_{iso} is the isothermal Mach number of the cold shell. The shell gas compression is:

$$\frac{\rho_{sh}}{\rho_{ISM}} = 170 V_{sh,2}^2 T_{ISM,4}^{-1}, \tag{11}$$

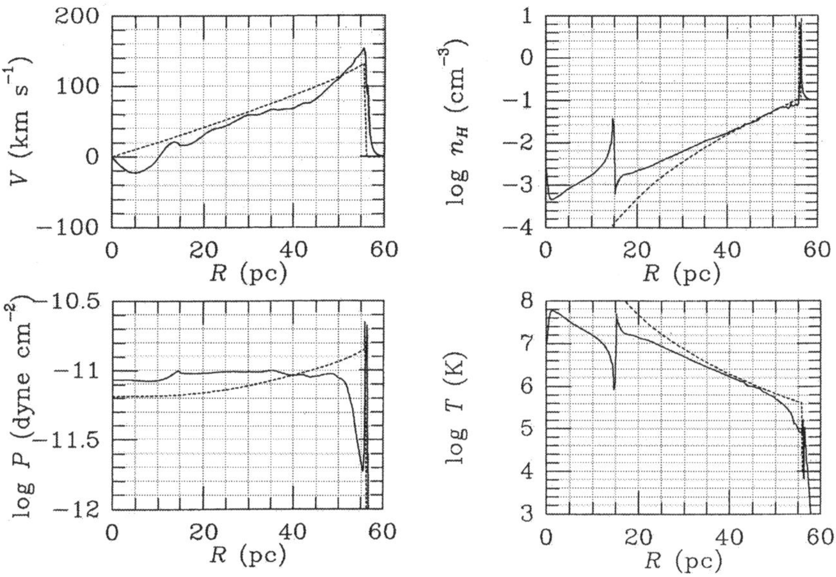

Fig. 2. Comparison of the proposed method (dash) with numerical simulation[2](solid) of basic gas flow characteristics for the SNR at transition stage. $E_{SN} = 0.931 \cdot 10^{51}$ ergs, $n_H = 0.1\ cm^{-3}$, age 170000 yrs.

where $V_{sh,2}$ is the shell velocity in units 100 km/s, $T_{ISM,4}$ is the shell temperature in $10^4 K$. The shell thickness is:

$$\Delta_{sh}(t) = \frac{M_{sh}(t)}{4\pi R_{sh}^2 \rho_{sh}} \sim \frac{R_{sh}}{3M_{iso}} \tag{12}$$

that is much less than the shell radius.

Our model was tested by comparison with the results of numerical simulations.[1,2,9] From Figs. 1, 2 it is seen that the proposed approximate analytical description represents the numerical results with high enough precision.

3. High energy signatures of transition stage

3.1. *X-ray emission*

The total X-ray luminosity L_x of the SNR can be calculated by integrating over the SNR volume and the surface brightness S by integrating along the line of sight inside the SNR:

$$L_x = \int_V \Lambda(T) n_e n_H dV \tag{13}$$

$$S = \frac{1}{4\pi} \int_{y_1}^{y_2} \Lambda(T) n_e n_H dy, \tag{14}$$

where the cooling function $\Lambda(T)$ is taken from Ref. 6. One of the features of transition stage is decline of the X-ray luminosity and flattening of the surface brightness profile of the SNR. It is explained by cooling of the essential part of the hot gas. The most visible falling is in soft range 0.1 keV $< E_x <$ 2.4 keV that is generated in the front region while in harder range that is generated in the inner and the hotter layers the relative drop of luminosity is lower. The evolution of the X-ray luminosity and the surface brightness profile can be seen at Fig. 3.

3.2. γ-ray emission from SNRs

In Ref. 12 it was shown that in the case of pulsar absence, the most promising mechanism for $\varepsilon_\gamma > 100$ MeV γ-ray production is inelastic interaction of relativistic protons with protons at rest resulting in the creation of pions and their consequent decay into γ-rays. The total energy of CRs in the SNR is $W_{CR} = \nu E_{SN}$, where ν is of order of 10^{-2} and the total number of CR is:

$$N_{tot} = \int N(\varepsilon) d\varepsilon = \frac{W_{cr}}{E_{cr}}, \tag{15}$$

where $E_{cr} \sim 10^9 eV$. The γ-ray luminosity equals to the rate of energy transformation from relativistic protons to neutral pions:[3,4,10]

$$L_\gamma = c n_N \int_{\varepsilon_{\min}}^{\infty} N(\varepsilon) \sigma_{pp}(\varepsilon) \bar{\varepsilon}_{\pi^o}(\varepsilon) d\varepsilon = \frac{c\overline{\sigma}_{pp}}{6} n_N W_{CR}, \tag{16}$$

where c is the velocity of light, $n_N = 1.4 n_H$ is the mean number density of target nuclei in the region of interaction, $\varepsilon_{\min} \approx 600$ MeV is the minimal proton kinetic energy of the effective pion creation (with the cross-section $\sigma_{pp}(\varepsilon)$ close to the mean value $\overline{\sigma}_{pp} = 3 \cdot 10^{-26}$ cm^2), $\bar{\varepsilon}_{\pi^o}(\varepsilon) = \varepsilon/6$ is the mean energy transformed into the pion.

The prominent feature of transition stage is the increase in the γ-ray luminosity of the SNR. For the case of uniform CR distribution inside the SNR at the end of adiabatic stage the total number of CR contained in the shell at the time t will be $N_{tot,sh}(t) = (3N_{tot}V(t))/4\pi R(t_{tr})^3$, where $V(t)$ is volume between $r(a_c(t), t_{tr})$ and $R(t_{tr})$. Energy of CRs in the cooling hot plasma increases in $(\rho_{shell}/\rho_{sw})^{1/3}$ times due to increasing of the frozen in magnetic field.[5] The compression of the shell also leads to a few orders

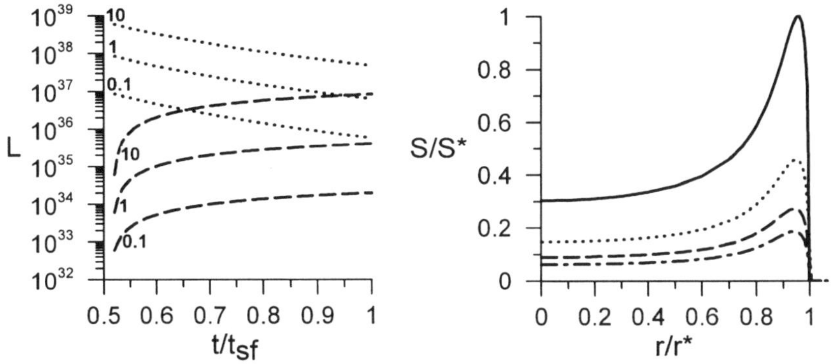

Fig. 3. Left - evolution of the SNR X-ray (dots) in the range > 0.1 keV and γ-ray (dash) in the range > 0.1 GeV luminosity (L is in erg/s) during transition stage for different number densities (n_H is in cm^{-3}). Right - evolution of the surface brightness profiles in X-rays during transition stage: $t = t_{tr} = 29000$ yrs (solid), $t = 37920$ yrs (dots), $t = 46840$ yrs (dash) $t = t_{sf} = 55760$ yrs (dash-dot). The explosion energy $E_{SN} = 10^{51}$ ergs, $n_H = 1\ cm^{-3}$, t_{sf} - the time of the shell formation, $S^* = S(t_{tr})$, $r^* = R(t)$.

gain in the number density of target nuclei that naturally leads to the increase in the γ-ray luminosity of the SNR (Fig. 3). At radiative stage the total number of CRs in the shell is constant but because of geometrical divergence of the shell, both the energy of CRs and the number density of targets decrease. Hence, the γ-ray luminosity should decrease as well.

4. Conclusion

In our work we developed the approximate analytical description of transition stage of the SNR evolution that is between adiabatic and radiative stages. This method generalizes the proposed earlier approximate description of adiabatic stage.[6] On its base we carried out calculations of basic high energy signatures of transition stage. These signatures include decline of the total X-ray luminosity of the SNR, flattening of its surface brightness profile and steep increase in the γ-ray flux with subsequent gentle decrease at radiative stage.

Acknowledgments

I.T. is grateful to the Direction of the International School of Cosmic Ray Astrophysics 2006 for support of his participance.

This work was supported by the Swiss National Science Foundation and the Swiss Agency for Development and Cooperation in the framework of the programme SCOPES - Scientific co-operation between Eastern Europe and Switzerland.

References

1. Blondin, J., Wright, E., Borkowski, K., Reynolds, S., Aph. J., 500, 342 (1998).
2. Cioffi, D.F., McKee, C.F., & Bertschinger, E., Aph. J., 334, 252 (1988).
3. Esposito, J.A. et al. Aph. J., 461, 820 (1996)
4. Drury, L.O., Aharonian, F.A., Volk, H.J. Astr. & Aph., 287, 959 (1994).
5. Hnatyk, B., Petruk, O. Cond.Mat.Phys., Vol.1, No.3(15), 655 (1998)
6. Hnatyk, B., Petruk, O. Astr. & Aph., 344, 295 (1999)
7. Jones, T.W. et al., PASP, Vol. 110, Issue 744, pp. 125-151. (1998)
8. Ostriker, J.P., & McKee, C.F., Rev.Mod. Phys., 60, 1 (1988).
9. Shelton, R.L. et al. Aph. J., 524, 192 (1999).
10. Berezinsky, V.S. et al. Astrophysics of cosmic rays. (Amsterdam, 1990).
11. Heavens, A.F. MNRAS, 1984, vol. 207, No 1, p. 1P-5P.
12. Sturner, S.J., Dermer, C.D. A&A, 1995, vol. 293, p. L17-L20.

THE NATURE OF DARK MATTER

Peter L. Biermann*[1,2,3], and Faustin Munyaneza†‡[1]

[1] *Max-Planck Institute for Radioastronomy, Bonn, Germany*
[2] *Department of Physics and Astronomy, University of Bonn, Germany,*
[3] *Department of Physics and Astronomy, University of Alabama, Tuscaloosa, AL, USA*

Dark matter has been recognized as an essential part of matter for over 70 years now, and many suggestions have been made, what it could be. Most of these ideas have centered on Cold Dark Matter, particles that are expected in extensions of standard particle physics, such as supersymmetry. Here we explore the concept that dark matter is sterile neutrinos, a concept that is commonly referred to as Warm Dark Matter. Such particles have keV masses, and decay over a very long time, much longer than the Hubble time. In their decay they produce X-ray photons which modify the ionization balance in the early universe, increasing the fraction of molecular Hydrogen, and thus help early star formation. Sterile neutrinos may also help to understand the baryon-asymmetry, the pulsar kicks, the early growth of black holes, the minimum mass of dwarf spheroidal galaxies, as well as the shape of dark matter halos. As soon as all these tests have been quantitative in its various parameters, we may focus on the creation mechanism of these particles, and could predict the strength of the sharp X-ray emission line, expected from any large dark matter assembly. A measurement of this X-ray emission line would be definitive proof for the existence of may be called weakly interacting neutrinos, or WINs.

Keywords: Dark matter, sterile neutrinos, galaxies, black hole physics

1. Dark Matter: Introduction

Since the pioneering works of Oort[1] and Zwicky,[2,3] we know that there is dark matter in the universe, matter that interacts gravitationally, but not measureably in any other way. Oort argued about the motion and density of stars perpendicular to the Galactic plane, and in this case, Oort's original hunch proved to be correct, the missing matter turned out to be low luminosity stars. Zwicky argued about the motions and densities of

*E-mail:plbiermann@mpifr-bonn.mpg.de
†E-mail:munyanez@mpifr-bonn.mpg.de
‡Humboldt Fellow

galaxies in clusters of galaxies, and to this day clusters of galaxies are prime arguments to determine dark matter, and its properties.

Based on the microwave back ground fluctuations[4] today we know that the universe is flat geometrically, i.e. the sum of the angles in a cosmic triangle is always 180 degrees, provided we do not pass too close to a black hole. This finding can be translated into stating that the sum of the mass and energy contributions to the critical density of the universe add up to unity, with about 0.04 in baryonic matter, about 0.20 in dark matter, and the rest in dark energy; we note that there is no consensus even where to find all the baryonic matter, but a good guess is warm to hot gas, such as found in groups and clusters of galaxies, and around early Hubble type galaxies.

There are many speculations of what dark matter is; we have three constraints:

1) It interacts almost exclusively by gravitation, and not measurably in any other way; 2) It does not participate in the nuclear reactions in the early universe; 3) It must be able to clump, to help form galaxies, and later clusters of galaxies, and the large scale structure.

Obviously, various extensions in particle physics theory, such as supersymmetry, all provide candidates, like the lightest supersymmetric particle.

Here we focus on the concept that it may be a "sterile neutrino", a right-handed neutrino, that interacts only weakly with other neutrinos, and otherwise only gravitationally. Such particles were first proposed by Pontecorvo[5] and later by Olive & Turner.[6] Sterile neutrinos were further proposed as dark matter candidates.[7] It was then shown how oscillations of normal neutrinos to sterile neutrinos could help explain the very large rectilinear velocities of some pulsars.[8]

Observationally the evidence comes from a variety of arguments: i) Dark matter in a halo like distribution is required to explain the stability of spiral galaxy disks;[9,10] ii) the flat rotation curves of galaxies[11]); and iii) the containment of hot gas in early Hubble type galaxies.[12] Dark matter is required to explain iv) the structure of clusters of galaxies;[13] v) structure formation, and the flat geometry of the universe.[4,14] We refer the reader to a recent review on dark matter.[15]

Therefore after more than 70 years we still face the question: "What is dark matter?"

2. Proposal

The existing proposals to explain dark matter mostly focus on very massive particles,[15] such as the lightest supersymmetric particle; all the experimental searches are sensitive for masses above GeV, usually far above such an energy. In the normal approach to structure formation, this implies a spectrum of dark matter clumps extending far down to globular cluster masses and below. It has been a difficulty for some time that there is no evidence for a large number of such entities near our Galaxy. The halo is clumpy in stars, but not so extremely clumpy. If, however, the mass of the dark matter particle were in the keV range, then the lowest mass clumps would be large enough to explain this lack. However, in this case the first star formation would be so extremely delayed[16] that there would be no explanation of the early reionization of the universe, between redshifts 11 and 6, as we now know for sure.[4,14] Therefore, the conundrum remained.

Here we explore the concept that the dark matter is indeed of a mass in the keV range, but can decay, and so produce in its decay a photon, which ionizes, so modifies the abundance of molecular Hydrogen, and so allows star formation to proceed early.[17,18] The specific model we explore is of "sterile neutrinos", right handed neutrinos, which interact only with normal, left-handed neutrinos, and with gravity. Such particles are commonly referred to as "Warm Dark Matter", as opposed to "Cold Dark Matter", those very massive particles. For most aspects of cosmology warm dark matter and cold dark matter predict the same; only at the small scales are they significantly different, and of course in their decay.

The mass range we explore is approximately 2 - 25 keV. These sterile neutrinos decay, with a very long lifetime, and in a first channel give three normal neutrinos, and in the second channel, a two-body decay, give a photon and a normal neutrino. The energy of this photon is almost exactly half the mass of the initial sterile neutrino.

What is important is to understand that such particles are not produced from any process in thermal equilibrium, and so their initial phase space distribution is far from thermal; all the current models for their distribution suggest that their momenta are sub-thermal. The measure of how much they are sub-thermal modifies the precise relationship between the dark matter particle mass and the minimum clump mass, which should be visible in the smallest pristine galaxies.

This also entails, that as Fermions they require a Fermi-Dirac distribution, as being far from equilibrium, this distribution implies a chemical potential.

Recent work by many others[19–30] has shown that these sterile neutrinos can be produced in the right amount to explain dark matter, could explain the baryon asymmetry,[31] explain the lack of power on small scales (as noted above), and could explain the dark matter distribution in galaxies.[32–34]

2.1. *Our recent work*

Pulsars are observed to reach linear space velocities of up to over 1000 km/s, and there are not many options how to explain this; one possibility is to do this through magnetic fields which become important in the explosion.[35–39] Another possibility is to do this through a conversion of active neutrinos which scatter with a mean free path of about ten cm, into sterile neutrinos, which no longer scatter. If this conversion produces a spatial and directional correlation between the sterile neutrinos and the structure of the highly magnetic and rotating core of the exploding star, then a small part of the momentum of the neutrinos can give an asymmetric momentum to the budding neutron star, ejecting it at a high velocity.[40] This then could explain such features as the guitar nebula, the bow shock around a high velocity pulsar. This latter model in one approximation requires a sterile neutrino in the mass range 2 to 20 keV. It is remarkable that this neutrino model requires magnetic fields in the upper range of the strengths predicted by the magneto-rotational mechanism to explode massive stars as supernovae.

It was also shown from SDSS data, that some quasars have supermassive black holes already at redshift 6.41, so 800 million years after the big bang.[41,42] We now know, that this is exactly when galaxies grow the fastest, from 500 to 900 million years after the big bang[43,44] Baryonic accretion has trouble feeding a normal black hole to this high mass, 3 10^9 solar masses so early after the big bang, if the growth were to start with stellar mass black holes.[45] So either the first black holes are around 10^4 to 10^6 solar masses, and there is not much evidence for this, or the early black holes grow from dark matter,[46–48] until they reach the critical minimum mass to be able to grow very fast and further from baryonic matter, which implies this mass range, 10^4 to 10^6 solar masses. This model in the isothermal approximation for galaxy structure implies a sterile neutrino in the mass range between 12 and 450 keV.

When Biermann and Kusenko met at Aspen meeting September 2005, it became apparent, that these two speculative approaches overlap, and so it seemed worth to pursue them further.

As noted above, structure formation arguments lead to an over-prediction in power at small scales in the dark matter distribution in the case of cold dark matter, and any attempt to solve this with warm dark matter delayed star formation unacceptably. We convinced ourselves that this was the key problem in reconciling warm dark matter (keV particles) with the requirements of large scale structure and reionization. We then showed that the decay of the sterile neutrino could increase the ionization, sufficiently to enhance the formation of molecular hydrogen, which in turn can provide catastrophic cooling early enough to allow star formation as early as required.[17,18] In our first simple calculation this happens at redshift 80. More refined calculations confirm, that the decay of sterile neutrinos helps increase the fraction of molecular Hydrogen, and so help star formation, as long as this is at redshifts larger than about 20.[49–51]

3. The tests

3.1. *Primordial magnetic fields*

In the decay a photon is produced, and this photon ionizes Hydrogen: at the first ionization an energetic electron is produced, which then ionizes much further, enhancing the rate of ionization by a factor of about 100. In the case, however, that there are primordial magnetic fields, this energetic electron could be caught up in wave-particle interaction, and gain energy rather than lose energy. As the cross section for ionization decreases with energy, the entire additional ionization by a factor of order 100 would be lost in this case, and so there basically would be no measurable effect from the dark matter decay. This gives a limit for the strength of the primordial magnetic field, given various models for the irregularity spectrum of the field: In all reasonable models this limit is of order a few to a few tens of picoGauss, recalibrated to today. Recent simulations matched to the magnetic field data of clusters and superclusters, give even more stringent limits, of picoGauss or less.[52]

It follows that primordial magnetic fields can not disturb the early ionization from the energetic photons, as a result of dark matter decay. It then also follows that the contribution of early magnetic fields from magnetic monopoles, or any other primordial mechanism, is correspondingly weak.[53]

Stars at all masses are clearly able to produce magnetic fields,[54–56] but the evolution and consequent dispersal are fastest for the massive stars, almost certainly the first stars. As the magnetic fields may help to drive the wind of these massive stars,[57] then the wind is just weakly super-Alfvénic, with Alfvénic Machnumbers of order a few. This implies that the massive stars and their winds already before the final supernova explosion may provide a magnetic field which is at order 10 percent equipartition of the environment; this magnetic field is highly structured. However, even these highly structured magnetic fields will also allow the first cosmic rays to be produced, and distributed, again with about 10 percent of equipartition of the environment.

However, the large scale structure and coherence of the cosmic magnetic fields clearly remain an unsolved problem.[58–60]

Therefore the first massive stars are critical for the early evolution of the universe: In addition to reionization, magnetic fields and cosmic rays, they provide the first heavy elements. These heavy elements allow in turn dust formation, which can be quite rapid (as seen, e.g., in SN 1987A, already just years after the explosion.[61] This then enhances the cooling in the dusty regions, allowing the next generation of stars to form much faster.

In combination everywhere one first massive star is formed, we can envisage a runaway in further star formation in its environment.

3.2. *Galaxies*

Galaxies merge, and simulations demonstrate that the inner dark matter distribution attains a power law in density, and a corresponding power law tail in the momentum distribution:[62–64] Here the central density distribution as a result of the merger is a divergent power law, as a result of energy flowing outwards and mass flowing inwards, rather akin to accretion disks[65,66] where angular momentum flows outwards and mass also flows inwards; in fact also in galaxy mergers angular momentum needs to be redistributed outwards as such mergers are almost never central.[67] This then leads to a local escape velocity converging with r to zero also towards zero, and so for fermions the Pauli limit is reached, giving rise to a cap in density, and so a dark matter star or a fermion ball is born;[46,47] this dark matter star can grow further by dark matter accretion. The physics of fermion balls at galactic centers has been studied in a series of papers.[68–74] For realistic models an integral over a temperature distribution is required, and a boundary condition has to be used to represent the surface of the dark matter star both in real space as in momentum space. This then allows the mass of this dark matter star to increase; such models resemble in their quantum statistics white dwarf stars or neutron stars; the Pauli pressure upholds the star. For fermions in the keV range the mass of the dark matter star has a mass range of a few thousand to a few million solar masses.

The first stellar black hole can then enter this configuration and eat the dark matter star from inside, taking particles from the low angular momentum phase space. With phase space continuously refilled through the turmoil of the galaxy merger in its abating stages, or in the next merger, the eating of the dark matter star from inside ends only when all the dark matter star has been eaten up.

Given a good description of the dark matter star boundary conditions in real and in momentum phase space,[46,47] and an observation of the stellar velocity dispersion close to the final black hole, but outside its immediate radial range of influence, we should be able to determine a limit to the dark matter particle mass. If the entire black hole in the Galactic Center has grown from dark matter alone, then we obtain a real number.

This concept suggests that it might be worthwhile to consider the smallest of all black holes in galactic centers. In a plot of black hole mass versus central stellar velocity dispersion σ there is a clump above the relation $M_{BH} \sim \sigma^4$, at low black hole masses,[75,76] suggesting that perhaps we reach a limiting relationship with a flatter slope for all those black holes which grow only from dark matter;[47] for a simple isothermal approach this flatter slope is found to be 3/2.

3.3. *Dwarf spheroidal galaxies*

All detected dwarf spheroidal galaxies fit a simple limiting relationship of a common dark matter mass of $5\,10^7$ solar masses,[34] suggesting that this is perhaps the smallest dark matter clump mass in the initial cosmological dark matter clump spectrum. This clump mass is of course a lower limit to the true original mass of the pristine dwarf spheroidal galaxy. Given a physical concept for the production of the dark matter particles in the early universe, we would have their initial momentum, probably subthermal, and so the connection between the dark matter particle mass and minimum clump mass is modified. This is very strong support for the Warm Dark Matter concept.

One intriguing aspect of dwarf spheroidal galaxies is that almost all of them show the effect of tidal distortion in their outer regions, and at least one of them has been distended to two, perhaps even three circumferential rings around our Galaxy.[32,33] To extend so far around our Galaxy must have taken many orbits, and so a some fraction of the age of our Galaxy. The simple observation that these streamers still exist separately, and can be distinguished in the sky, after many rotations around our Galaxy, implies that the dark matter halo is extremely smooth, and also nearly spherically symmetric. Given that the stellar halo is quite clumpy this implies once more that the dark matter is much more massive than the baryonic matter in our halo.

3.4. *Lyman alpha forest*

In the early structure formation the large number of linear perturbations in density do not lead to galaxies, but just too small enhancements of Hydrogen density, visible in absorption against a background quasar. This so-called Lyman alpha forest tests the section of the perturbation scales which is linear and so much easier to understand, and it should in principle allow a test for the smallest clumps.[77] Unfortunately, systematics make this test still difficult, and with the expected sub-thermal phase space distribution of the dark matter particles we may lack yet the sensitivity to determine the mass of the smallest clumps.

3.5. *The X-ray test*

When the sterile neutrinos decay, they give off a photon with almost exactly half their mass in energy. Our nearby dwarf spheroidal galaxies, our own inner Galaxy, nearby massive galaxies like M31, the next clusters of galaxies like the Virgo cluster, and other clusters further away, all should show a sharp X-ray emission line.[19,78–80]

The universal X-ray background should show such a sharp emission line as a wedge, integrating to high redshift.

With major effort this line or wedge be detectable with the current Japanese, American or European X-ray satellites: Large field high spectral resolution spectroscopy is required.

4. Outlook

The potential of these right handed neutrinos is impressive, but in all cases we have argued, there is a way out, in each case there is an alternative way to interpret the data set. E.g., for the pulsar kick with the help of neutrinos strong magnetic fields are required, but the MHD simulations suggest that perhaps magnetic fields can do it by themselves, even without the weakly interacting neutrinos.[35,38,39] The dwarf spheroidal galaxies can in some models be explained without any dark matter at all.[81,82] The early growth of black holes can also be fueled by other black holes, as long as here are enough in number and their angular momentum can be removed. So many alternatives may replace the sterile neutrino concept.

However, the right handed, sterile neutrinos weakly interacting with the normal left handed neutrinos provide a unifying simple hypothesis, which offers a unique explanation of a large number of phenomena, so by Occam's razor, it seems quite convincing at present.[83] So, given what sterile neutrinos may effect, we may have to call them Weakly Interacting Neutrinos, or soon WINs.

5. Acknowledgements

The authors wish to thank first and foremost Alex Kusenko, an indefatigable partner in all explorations of warm dark matter; he played a key role in working out the science reported here. The authors would also like to acknowledge fruitful discussions with Kevork Abazajian, Gennadi Bisnovatyi-Kogan, Gerry Gilmore, Phil Kronberg, Pavel Kroupa, Sergei Moiseenko, Biman Nath, Mikhail Shaposhnikov, Simon Vidrih, Tomaz Zwitter, and many others.

Support for PLB is coming from the AUGER membership and theory grant 05 CU 5PD 1/2 via DESY/BMBF. Support for FM is coming from the Humboldt Foundation.

References

1. J.H. Oort, *Bull. Astron. Inst. Netherl.* , **VI**, 249 (1932)
2. F. Zwicky, *Helv. Phys. Acta* **6**, 110 (1933)
3. F. Zwicky, *Astrophys. J.* **86**, 217 (1937)
4. D. N. Spergel et al., 2006, *Astrophys. J.* (submitted); astro-ph/0603449
5. B. Pontecorvo,*J.E.T.P* **53**, 1717 (1967)
6. K.A. Olive and M.S. Turner, Phys. Rev. D **25**, 213 (1982)
7. S. Dodelson and L. M. Widrow, Phys. Rev. Lett. **72**, 17 (1994)
8. A. Kusenko and G. Segrè, *Phys. Lett. B* **396**, 197 (1997);
9. J.P. Ostriker and P.J.E Peebles, *Astrophys. J.* **186**, 467 - (1973)
10. J.P. Ostriker, P.J.E. Peebles, and A. Yahil, *Astrophys. J. Letters* **193**, L1 (1974)
11. V.C. Rubin, N. Thonnard and W.K. Ford Jr., *Astrophys. J.* **238**, 471 (1980)
12. P.L Biermann, P.P. Kronberg and B.F. Madore *Astrophys. J. Letters* **256**, L37 (1982)
13. T. A. Enßlin, P. L. Biermann, P. P. Kronberg, and Xiang P. Wu, *Astrophys. J.* **477**, 560 (1997)
14. D. N. Spergel *et al.* [WMAP Collaboration], Astrophys. J. Suppl. **148**, 175 (2003).
15. G. Bertone, D. Hooper and J. Silk, *Physics Reports*, **405**, 279 (2005)
16. N. Yoshida, A. Sokasian, L. Hernquist and V. Springel,*Astrophys. J. Letters* **591**, L1 (2003)

17. P.L. Biermann and A. Kusenko, 2006, *Phys. Rev. Letters* **96**, 091301 (2006)
18. J. Stasielak, P.L. Biermann and A. Kusenko, in press, *Astrophys. J.* (2006), astro-ph/0606435
19. K. Abazajian, G.M. Fuller and M. Patel *Phys. Rev. D* **64**, 023501 (2001)
20. K. Abazajian, *Phys. Rev.* **D 72**, id. 063004 (2005)
21. K. Abazajian and S.M. Koushiappas, *Phys. Rev.* **D** (submitted, 2006), astro-ph/0605271
22. K. Abazajian, *Phys. Rev.* **D 73**, 063513 (2006),
23. K. Abazajian, *Phys. Rev.* **D 73**, id. 063506 (2006),
24. A. Boyarsky,A. Neronov, O. Ruchayskiy, M. Shaposhnikov, and I. Tkachev, astro-ph/0603660 (2006)
25. A. Boyarsky, A. Neronov, O. Ruchayskiy, and M. Shaposhnikov, astro-ph/0603368 (2006)
26. A. Boyarsky, A. Neronov, O. Ruchayskiy and M. Shaposhnikov, *Month. Not. Roy. Astr. Soc.* in press, (2006), astro-ph/0512509
27. A.D. Dolgov and S.H. Hansen, *Astr. Part. Phys.* **16** 339 (2002)
28. T. Asaka, and M. Shaposhnikov, *Phys. Lett. B* **620**, 17 (2005)
29. T. Asaka, St. Blanchet and M. Shaposhnikov, *Phys. Lett. B* **631**, 151 (2005)
30. T. Asaka, M. Shaposhnikov and A. Kusenko *Phys. Lett. B* **638**, 401 (2006)
31. E. K. Akhmedov, V. A. Rubakov and A. Y. Smirnov, *Phys. Rev. Lett.* **81**, 1359 (1998).
32. V. Belokurov, *et al.*, *Astrophys. J. Letters* **642**, L137 - L140 (2006), astro-ph/0605025)
33. M. Fellhauer *et al.*, (2006), astro-ph/0605026
34. G. Gilmore, et al., UCLA Dark Matter 2006 conference, March 2006. In press in *Nuclear Physics* **B**, (2006); astro-ph/0608528
35. G.S. Bisnovatyi-Kogan, *Sov. Astron. Astron. J.*, **14**, 652 (1971), orig. *Astron. Zh* **47**, 813 (1970).
36. G.S. Bisnovatyi-Kogan, *Astron. Astrophys. Trans.* **3** 287 (1993)
37. G.S. Bisnovatyi-Kogan, *AIP Conf. Proc.* **366** 38 (1995)
38. N.V. Ardeljan, G.S. Bisnovatyi-Kogan, and S.G. Moiseenko, *Month. Not. Roy. Astr. Soc.* **359**, 333 (2005)
39. S. G. Moiseenko, G.S. Bisnovatyi-Kogan and N.V. Ardeljan, *Mem. Soc. Astron. Ital.* **76**, 575 (2005)
40. A. Kusenko, *Int. J. Mod. Phys. D* **13**, 2065 (2004).
41. X. Fan et al., *Astron. J.* **122**, 2833 - 2849 (2001)
42. C.J. Willott, R.J. McLure, M.J. Jarvis, *Astrophys. J. Letters* **587**, L15 (2003)
43. R. J. Bouwens and G.D. Illingworth, *Nature* **443**, 189 (2006)
44. M. Iye et al., *Nature* **443**, 186 (2006)
45. Y. P. Wang, and P. L. Biermann, *Astron. & Astroph.* **334**, 87 (1998)
46. F. Munyaneza and P.L. Biermann, *Astron. & Astroph.* **436**, 805 (2005)
47. F. Munyaneza and P.L. Biermann, *Astron. & Astroph. Letters* **458**, L9 (2006)
48. F. Munyaneza and P.L. Biermann, *Proceedings of the Galactic Center Workshop* Bad Hönnef, April 18-22, 2006 eds. R. Schödel et al., *Journal of Physics: Conference series* **54**, 456 (2006)
49. M. Mapelli and A. Ferrara, Mon. Not. Roy. Astron. Soc. **364**, 2 (2005).

50. M. Mapelli, A. Ferrara and E. Pierpaoli, *Month. Not. Roy. Astr. Soc.* **369**, 1719 (2006)
51. E. Ripamonto, M. Mapelli and A. Ferrara, *Month. Not. Roy. Astr. Soc.* in press (2006); astro-ph/0606483
52. K. Dolag, D. Grasso, V. Springel, and I. Tkachev, *J. of Cosm. & Astrop. Phys.* **01**, 009 (2005)
53. S.D. Wick, T. W. Kephart, T.J. Weiler and P.L. Biermann, *Astropart. Phys.* **18**, 663 (2003)
54. L. Biermann, *Z. f. Naturf.* **5a**, 65 - 71 (1950)
55. L. Biermann and A. Schlüter, *Phys. Rev.* **82**, 863 (1951)
56. J. Silk and M. Langer, *Month. Not. Roy. Astr. Soc.* **371**, 444 (2006)
57. H. Seemann and P.L. Biermann, *Astron. & Astroph.* **327**, 273 (1997),
58. R.M. Kulsrud, *Annual Rev. of Astron. & Astrophys.* **37**, 37 (1999)
59. P. L. Biermann and C.F. Galea, in *the 9th course of the Chalonge School on Astrofundamental Physics: "The Early Universe and The Cosmic Microwave Background: Theory and Observations"*; Eds. N.G. Sanchez & Y.N. Parijski, Kluwer, 471 (2003),
60. P.L. Biermann and P.P. Kronberg, in *Proc. Pusan conference*, August 2004, Ed. D. Ryu, *Journal of the Korean Astronomical Society*, **37**, 527 (2004)
61. P.L. Biermann et al. *Astron. & Astroph. Letters* **236**, L17 (1990)
62. J.F. Navarro, C.S. Frenk and S.D.M White, *Astrophys. J.* **490**, 493 (1997)
63. B. Moore et al., *Astrophys. J. Letters* **524**, L19 (1999)
64. A. Klypin, H.S. Zhao and R.S. Somerville, *Astrophys. J.* **573**, 597 (2002)
65. R. Lüst, *Zeitschr. f. Naturf.* **7a** 87 (1952)
66. R. Lüst and A. Schlüter, *Z. f. Astroph.* **38**, 190 (1955)
67. A. Toomre and J. Toomre, *Astrophys. J.* **178**, 623 (1972)
68. R.D. Viollier, *Prog. Part. Nucl. Phys.* **32**, 51 (1994)
69. D. Tsiklauri and R.D. Viollier, *Astrophys. J.* **500**, 591 (1998)
70. F. Munyaneza, D. Tsiklauri and R.D. Viollier, *Astrophys. J.* **509**, L105 (1998)
71. F. Munyaneza, D. Tsiklauri and R.D. Viollier, *Astrophys. J.* **526**, 744 (1999)
72. N. Bilić, F. Munyaneza and R.D. Viollier, 1999, *Phys. Rev.* **D. 59**, 024003 (1999)
73. F. Munyaneza and R.D. Viollier, *Astrophys. J.* **564**, 274 (2002)
74. N. Bilić, F. Munyaneza, G. Tupper and R.D. Viollier, *Prog. Part. Nucl. Phys.* **48**, 291 (2002)
75. A.J. Barth, J.E. Greene and L.C. Ho, *Astrophys. J. Letters* **619**, L151 (2005)
76. J.E. Greene, A.J. Barth and L.C. Ho, *New Astron. Rev.* **50**, 739 (2006)
77. M. Viel, J. Lesgourgues, M.G. Haehnelt, S. Mattares and A. Riotto, *Phys. Rev. Letters* **97**, 071301 (2006)
78. S. Riemer-Sørensen, St. H. Hansen, and K. Pedersen, *Astrophys. J. Letters* **644**, L33 (2006)
79. S. Riemer-Sørensen, K. Pedersen, St. H. Hansen, and H. Dahle, astro-ph/0610034, (2006)
80. C.R. Watson, J.F. Beacom, H. Yüksel and T.P. Walker, *Phys. Rev.* **D74**, 033009 (2006)

81. M. Metz, P. Kroupa and H. Jerjen, *Month. Not. Roy. Astr. Soc.* (in press, 2006); astro-ph/0610933
82. S.T. Sohn et al., submitted (2006); astro-ph/0608151
83. A. Kusenko, *Phys. Rev. Letters* **97**, 241301 (2006)

cosmic rays

PARTICLE ACCELERATION AND PROPAGATION IN THE GALAXY

VLADIMIR S. PTUSKIN
Institute of Terrestrial Magnetism, Ionosphere and Radiowave Propagation (IZMIRAN),
Troitsk, Moscow region 142190, Russia
E-mail: vptuskin@izmiran.ru

The processes of cosmic ray acceleration and transport in the Galaxy are briefly discussed.

Keywords: cosmic rays; supernova remnants; interstellar medium.

1. Introduction.

Our Galaxy is filled with cosmic rays - a relativistic gas of high-energy protons, electrons, and heavy nuclei. The major portion of these particles was accelerated in supernova remnants and is wondering for about 10^8 yr before exit to the intergalactic space. The particles with energies larger than $10^{18}-10^{19}$ eV have an extragalactic origin. High-energy particles are also an important and distinguishing feature of radio galaxies, quasars, and active galactic nuclei. The direct measurement by space and balloon experiments of their charge and mass composition and energy spectra provide information on the source regions within our Galaxy, on injection and acceleration processes, and offer a steadily increasing understanding of cosmic-ray transport through interstellar space. Observations from radio, gamma-ray, and X-ray astronomy define the distribution of energetic particles throughout our Galaxy and establish their presence in extragalactic sources. The interpretation of these observations by allied scientific disciplines is significantly aided by the detailed study of cosmic rays near Earth while our understanding of the sources and of the distribution of galactic cosmic rays is strongly dependent on the data from these other fields.

2. Diffusion

The motion of cosmic ray particles with energies up to $E \sim 10^{17}$ eV in galactic magnetic fields is usually described as diffusion [1]. The diffusion model serves as a basis for the interpretation of data on the spectrum, composition, and anisotropy of cosmic rays. A good fit to these data supported by the radio-astronomical and the gamma-ray observations allows one to determine the parameters of cosmic ray propagation model. The procedure requires the solution of transport equations for all cosmic ray species at the given source distribution and the given boundary conditions. The transport equation describes particle diffusion, convection, and energy changes which include the energy losses and possible distributed reacceleration by the interstellar turbulence. The high abundance of rare in nature elements and isotopes ^{2}H, ^{3}He, Li, Be, B and others are observed in cosmic rays. These secondary nuclei are produced in a course of nuclear fragmentation of primary energetic nuclei in the interstallar gas. The cosmic rays traverse on average about 10 g/cm^2 at the energy $\sim$ 1 GeV/nucleon where the maximum ratio of secondary to primary nuclei is observed.

The modeling of cosmic ray propagation gives the following set of parameters of the galactic diffusion model: the total power of cosmic ray sources in the Galaxy is $Q_{cr} = 5 \times 10^{40}$ erg/s (this comprises about 15% of the kinetic energy of supernova explosions), The height of the galactic halo is $H = 4$ kpc (or larger in the model with galactic wind). According to [2] the value of cosmic ray diffusion coefficient in two basic versions of the diffusion model is

$$D = 2.2 \times 10^{28} \beta \, (R/R_0)^{0.6} \, \text{cm}^2/\text{s at } R > R_0 = 3\text{GV}, \; D \sim \beta^{-2} \text{at } R < R_0 \tag{1}$$

in the plain diffusion model;

$$D = 5.2 \times 10^{28} \beta \, (R/R_0)^{0.34} \, \text{cm}^2/\text{s at all } R \tag{2}$$

in the diffusion model with distributed stochastic reacceleration in the interstellar medium by the mhd waves with Alfven velocity $V_a = 36$ km/s. Here $R = pc/Z$ is the particle magnetic rigidity, p is momentum, Z is the charge, and v is the particle velocity, $\beta = v/c$.

Both versions are not free of difficulties and need improvement. The strong energy dependence of diffusion in the plain diffusion model (1) leads to anisotropy that considerably exceeds the observed value at $\sim 10^{14}$ eV, see Section 2 below. On the other hand the model with reacceleration underestimates the flux of secondary antiprotons in cosmic rays. It is also important that the observed cosmic ray spectrum $\sim E^{-2.7}$ at $E > 30$ GeV/n implies the source spectrum $\sim E^{-2.1}$ in the version (1) and $E^{-2.36}$ in the version (2). The observations of gamma-rays from the supernova remnants and the modern theory of particle acceleration give the particle spectrum close to E^{-2}; thus the version (1) is preferable.

On the "microscopic level" the diffusion of cosmic rays results from the particle scattering on random MHD waves and discontinuities. The effective "collision integral" for charged energetic particles moving in a magnetic field with small random fluctuations $\delta B \ll B$ can be taken from the standard quasi-linear theory of plasma turbulence [3]. The wave-particle interaction is of resonant character so that an energetic particle is predominantly scattered by the irregularities of magnetic field that have the projection of wave vector on the average magnetic field direction equal to $k_{\parallel} = \pm s/(r_{\rm g}\mu)$, where μ is the particle pitch angle. The integers $s = 0, 1, 2...$ correspond to the cyclotron resonances of different orders. The efficiency of scattering depends on the polarization of the waves and on their distribution in **k**-space. The first-order resonance $s = 1$ is the most important for the isotropic and also for the one-dimensional distribution of random MHD waves along the average magnetic field. In some cases – for calculation of scattering at small μ and for calculation of perpendicular diffusion – the broadening of resonances and magnetic mirroring effects should be taken into account. The resulting spatial diffusion is strongly anisotropic locally and goes predominantly along the magnetic field lines. However, strong fluctuations of magnetic field on the large scale $L \sim 100$ pc where the strength of random field is several times higher than the average field strength, lead to the isotropization of global cosmic ray diffusion in the Galaxy. The rigorous consideration of this effect is not trivial, since the field is almost static and the strictly one-dimensional diffusion along the magnetic field lines does not lead to non-zero diffusion perpendicular to **B**, see [4,5].

With reference to the detailed reviews of the theory of cosmic ray diffusion [1,6] the diffusion coefficient of cosmic rays at $r_{\rm g} < L$ can be roughly estimated as $D \approx (\delta B_{\rm res}/B)^{-2} v r_{\rm g}/3$, where $\delta B_{\rm res}$ is the amplitude of random field at the resonant wave number $k_{\rm res} = 1/r_g$. The spectral energy density of interstellar turbulence has a power law form $w(k)dk \sim k^{-2+a}dk$, $a = 1/3$ in a wide range of wave numbers $1/(10^{20}\ \text{cm}) < k < 1/(10^{8}\ \text{cm})$, see [7], and the strength of random field at the main scale is $\delta B \approx 5\ \mu$G. It gives an estimate of the diffusion coefficient $D \approx 2 \times 10^{27} \beta R_{\rm GV}^{1/3}\ \text{cm}^2/\text{s}$ for all cosmic ray particles with magnetic rigidities $R < 10^8$ GV in a fair agreement with the empirical diffusion model (the version with distributed reacceleration). The scaling of diffusion $D \sim R^{1/3}$ is determined by the value of exponent $a = 1/3$ typical for Kolmogorov spectrum. Theoretically [8] the Kolmogorov type spectrum might refer only to some part of the MHD turbulence that includes the (Alfvenic) structures strongly elongated along the magnetic field direction and that are not able to provide the significant scattering and needed diffusion of cosmic rays. In parallel the more isotropic (fast magnetosonic) part of the turbulence with a smaller value of random field at the main scale and with the exponent $a = 1/2$ typical for the Kraichnan type spectrum may exist in the interstellar medium [9]. The Kraichnan spectrum gives the scaling $D \sim R^{1/2}$ that is close to the high energy asymptotic of the diffusion coefficient obtained in the "plain diffusion" version of the empirical propagation model. Thus the approach based on the kinetic theory gives a proper estimate of the diffusion coefficient and predicts a power law dependence of diffusion on magnetic rigidity but the ultimate determination of the diffusion coefficient should be done with the help of the empirical model of cosmic ray propagation in the Galaxy.

3. Supernova remnants

Energetically, supernovae are the most probable cosmic ray sources in the Galaxy [10,11]. About 10 to 20% of the kinetic energy of SN ejecta is needed to maintain the observed cosmic ray energy density $w_{cr} \sim 1.5$ eV/cm^3. It is assumed here that the kinetic energy of a SN explosion is 10^{51} erg and the supernova rate in the Galaxy is 1 every 30 years.

The clear evidence for particle acceleration in supernova remnants is presented by the observations of nonthermal radio, X-ray, and gamma-ray radiation. The data on synchrotron radio emission testifies the presence of electrons with energies 50 MeV - 30 GeV in such supernova remnants as Cas A, IC 433 and Cygnus Loop and many others, e.g. [12]. In the case of Cas A, this synchrotron emission was detected in the infrared waveband that proved the presence of electrons with energies up to about 2×10^{11} eV [13]. The detection of non-thermal X-rays radiation with a characteristic power law tail at energies up to $\sim$ 10 keV from the bright rims in $\sim$ 10 young supernova remnants including SN1006, Cas A, RXJ 1713.7-3946, RCW 86, G266.2-1.2 and some others is explained as the synchrotron emission of very high energy electrons with energies up to $10^{13} - 10^{14}$ eV, see [14] for review. The inverse Compton scattering of background photons by electrons with such high energies and the production of gamma-rays generated via $\pi^0 \rightarrow 2\gamma$ channel by protons and nuclei with energies up to 10^{14} eV/n interacting with background gas are the mechanisms of the emission of TeV gamma rays detected from about 10 SNRs (some observed gamma-ray sources are not reliably identified yet, see [15. In two cases of shell supernova remnants RX J1713.7-3946 [16] and SNR RX J0852.0-4622/Vela Jr [17,18] the TeV gamma-ray images with detailed angular resolution were obtained. The spatial distribution of nonthermal emission in all available energy range testifies that the acceleration of particles in the shell-type supernova remnants goes in the immediate shock region.

The particle acceleration by the outward propagating shock which results from the supernova explosion and propagates in the interstellar medium or in the wind of progenitor star is confirmed by the data on cosmic ray composition. After correcting for atomic selection which depends on the elements first ionization potential or vilotility (the ability to form stable compaunds and structures such as dust grains), the current propagation models yield a composition of accelerated material similar to the solar photosphere and to the local interstellar medium composition. The popular scenarios include the acceleration of ions and grains in the partly ionised interstellar medium [19,20] and the acceleration of material in hot superbables with high supernova rate [21,22]. The relatively high and close-to-the-solar value of the ratio ^{59}Co/^{56}Fe testifies that the major part of originally synthesized ^{59}Ni has decayed by the K-capture of an orbital electron into ^{59}Co before the acceleration started, see [23]. This means that the delay between synthesis of this material and the acceleration is larger than 10^5 yr.

The diffusive shock acceleration by supernova blast waves is the principal mechanism of cosmic ray acceleration in the Galaxy, see [11,24] for review. The acceleration of a fast particle diffusing near the shock front is a version of the first order Fermi acceleration. Fast particles are scattered by the inhomogeneities of magnetic field frozen into the background plasma and gain energy in a process of multiple crossing the shock - the region where the flow is converging. The characteristic time of particle acceleration by the shock with velocity u_{sh} can be estimated as D/u_{sh}^2, where D is the characteristic coefficient of cosmic ray diffusion which should proceed both upstream and downstream of the shock. The distribution of accelerated particles on momentum has a power law form $f(p) \sim p^{-3r/(r-1)}$, where r is the gas compression ratio in the shock. Recall that the cosmic ray intensity $I(E)$ is related to the distribution function $f(p)$ by the equation $f(p)p^2 = I(E)$. The compression in strong shocks is $r = 4$ and hence the spectrum of accelerated test particles is $f(p) \sim p^{-4}$ that is close to the required source spectrum of galactic cosmic rays.

This result is valid in the case of a step-like profile of the flow velocity at the shock or, more precisely, when the characteristic thickness of the shock L_{sh} is relatively small: $L_{sh} \ll D/u_{sh}$. The large density of energetic particles in the vicinity of high Mach number shock where the acceleration occurs causes the modification of gas flow through the action of energetic particle pressure and initiates plasma instabilities produced by the current of energetic particles. The modification of the flow by cosmic ray pressure leads to the violation of the thin shock approximation and results in a concave particle spectrum which deviates from a simple power-law form derived in a test particle approximation. Asymptotically at very high energies, the spectrum of accelerated particles may flatten out to $p^{-3.5}$ instead of p^{-4}.

The dependence of diffusion on energy determines the maximum energy that particles can gain in the process of acceleration. The necessary condition of efficient acceleration at the shock is $D(E) \leq 0.1 u_{sh} R_{sh}$, where R_{sh} is the radius of spherical shock (the value of the numerical factor 0.1 is approximate here). Notice that a typical supernova burst with kinetic energy of ejecta $W_{sn} = W_{51} 10^{51}$ erg in the interstellar gas with number density n_0 cm^{-3} gives the maximum value of the product $0.1 u_{sh} R_{sh} \sim 10^{27} (W_{51}/n_0)^{0.4}$ cm^2/s at the end of the free expansion stage of supernova remnant evolution when this product reaches its maximum value. At the same time, the typical value of cosmic ray diffusion coefficient in the Galaxy is close to 3×10^{28} cm^2/s at few GeV/n and is risen with energy. Thus the necessary condition of acceleration cannot be fulfilled for relativistic particles unless their diffusion coefficient is anomalously small in the vicinity of the shock.

The required slow diffusion in the shock precursor is selfconsistently produced by the streaming instability of accelerating particles that leads to the enhenced level of magnetohydrodynamic turbulence and the correspondent small diffusion coefficient. The theory of weak turbulence predicts strong amplification of random magnetic field δB at the shocks propagating with high Mach numbers [25] but it can not adequatly describe the amplification to the values comparable to the background interstellar field $B \approx 5 \times 10^{-6}$ G . Assuming the critical value for random field $\delta B \sim B$, one gets the so called Bohm value $D_B = v r_g/3 \approx 6 \times 10^{21} \beta R_m$ cm^2/s for the diffusion coefficient that is the lower bound for particle diffusion along the magnetic field (here v is the particle velocity, $\beta = v/c$, $r_g = pc/(ZeB)$ is the Larmor radius of a particle with momentum p and charge Ze, and $R_m = pc/Z$ is the particle magnetic rigidity in units GV). The Bohm diffusion allows particle acceleration to the maximum energy $E_{max} \sim 2 \times 10^{14} Z \,(W_{51}/n_0)^{0.4}$ eV at the time of transition from the ejecta-dominated stage to the stage of adiabatic evolution of supernova remnants. The assumption about Bohm diffusion in the vicinity of strong shocks was commonly used in the simulations of cosmic ray acceleration in supernova remnants. The recent study of strong streaming instability [26-28] proved that the amplified field may considerably exceed the initial magnetic field that allowing the shock acceleration up to more than $E_{\max} \sim 10^{16}$Z eV. This theoretical result is in agreement with the observed sharp structure of nonthermal X-ray rims, which is probably determined by the electron energy losses. The derived strength of amplified field in young supernova remnants Cas A, SN 1006, Tycho, RCW 86, Kepler, RX J1713.7-3946 considerably exceeds the interstellar magnetic field and reaches $(3-5) \times 10^{-4}$ G in historic supernova remnants [29,30].

4. Knee and above

The spectrum of very high energy cosmic rays with energies $E > 10^{12}$ eV is well described by a power law with three features that may be identified in such spectrum: a knee (steepening) at 3×10^{15} eV; a less pronounced second knee at about 10^{18} eV; and a flattening, "ankle", at $3 \times 10^{18}...10^{19}$ eV, see reviews [31-33]. The particles above the ankle are thought to be of extragalactic origin and their sources are associated with active galactic nuclei, gamma-ray bursts, interacting galaxies, etc [34] whereas the flux of galactic cosmic rays eventually dissolves in an extragalactic background at these highest energies. A fourth spectral feature at energies above 5×10^{19} eV, the Greisen-Zatsepin-Kuzmin cutoff, has been predicted to exist as the result of cosmic ray interaction with the cosmic microwave background and has been observed in the HiRes experiment [35].

The modern theory of diffusive shock acceleration where cosmic ray streaming instability creates very strong random magnetic field in the shock precursor gives the estimate of maximum particle energy $E_m \sim 10^{17} Z$ eV (Z is the particle charge) in first few weeks after the supernova explosion, and the observations of synchrotron X-ray and TeV gamma-ray emission from supernova remnants "directly" confirm the efficient particle acceleration up to energies $\sim 10^{14}$ eV in SNRs at the age $\sim 10^3$ yr. The theoretical consideration [36] showed that E_m decreases with a SNR age when the dissipation processes for the mhd turbulence are taken into consideration. The correspondingly calculated average source spectrum of accelerated ions injected in the interstellar space during the lifetime of a SNR has a pronounced break at about knee position $3 \times 10^{15} Z$ eV that marks the transition from the acceleration at the ejecta-dominated to the adiabatic (Sedov) evolution of SNR shocks, which accelerate cosmic rays. It might explain the knee at

3×10^{15} eV in the overall spectrum.

The knee can in principle be explained by an effect not related to acceleration itself. The escape probability from the Galaxy increases with energy and it may cause the knee, even if the source spectrum follows a single power law. For example, the knee might occur as a result of interplay between the diffusion of cosmic rays along magnetic field lines and the drift (Hall diffusion) perpendicular to the average, predominantly azimuthal, galactic magnetic field [37,38]. In any astrophysical scenario, the breaks and cutoffs in the spectra of ions with different charges should occur at the same magnetic rigidity, i.e. at the same ratio E/Z for the ultrarelativistic nuclei. Over the last decade, a set of experimental data in the energy range $10^{14} - 10^{16}$ eV has been collected. They yield a rather consistent picture and confirm the rise in average mass of the primary cosmic particles when passing the region of the knee at 3×10^{15} eV as it is expected if different ions experience the steepening at the same magnetic rigidity. The different models of the knee and their correspondence to the observational data were discussed in [39].

There is an obvious problem with the interpretation of the second knee in the cosmic ray spectrum, see e.g. [40]. The natural assumption that each individual ion has a spectrum break at the same ratio E/Z for all species and that the break in the spectrum of iron nuclei ($Z = 26$) expected at $\sim 8 \times 10^{16}$ eV explains the second knee in the all-particle spectrum does not agree with the observed position of the second knee at $5 \times 10^{17} - 10^{18}$ eV. To fit the overall spectrum at energies from about 10^{16} eV to 10^{18} eV, some supplementary cosmic ray component is needed in addition to the main galactic component. In principle, it can be caused by the dispersion of supernova parameters including the rare extremely energetic supernova events (the hypernova, the GRBs) [41]. Also, the required extension of the source spectrum could be obtained in the two-stage model where the individual supernova remnants accelerate particles up to $\sim 3 \times 10^{15}$ eV and the subsequent energy gain is due to the collective reacceleration on many shocks produced by other supernovae in the same OB star association [42]. Another option is the cosmic ray acceleration or reacceleration in the hypothetical Galactic wind, - by termination shock [43] or by traveling shocks [44].

Alternatively, the transition to extragalactic cosmic rays may happen at lower energies than thought traditionally, namely, at $\sim 5 \times 10^{17}$ eV [45]. The steep extragalactic source spectrum $\sim E^{-2.7}$ is required in this case and the observed cosmic ray spectrum at 5×10^{17} to 5×10^{19} eV with its characteristic dip (the region between second knee and the ankle) can be explained as the spectrum of extragalactic protons which interact with the microwave background and loose energy via electron-positron pair production. The comprehensive analysis of this scenario was given in [46]. The topic of extragalactic cosmic rays is beyond the scopes of our article.

Acknowledgments

The author is grateful to the organizers of the Erice School of Cosmic-Ray Astrophysics for hospitality and sponsorship.

References

1. V. S. Berezinskii et al., *Astrophysics of Cosmic Rays* (North Holland, Amsterdam, 1990).
2. V. S. Ptuskin et al., *ApJ* **642**, 902 (2006).
3. C. F. Kennel, F. Engelmann, *Phys. Fluids* **9**, 2377 (1066).
4. L. G. Chuvilgin, V. S. Ptuskin, *A&A* **279**, 278 (1993).
5. F. Casse et al., *Phys. Rev. D* **65**, 3002 (2002).
6. I. N. Toptygin, *Cosmic Rays in Interplanetary Magnetic Fields* (D. Reidel Publ. Co., Dordrecht, 1985).
7. B. G. Elmegreen, J. Scalo, *Ann. Rev. Astron. Astrophys.* **42**, 211 (2004).
8. P. Goldreich, S. Sridhar, *ApJ* **438**, 763 (1995).
9. H. Yan, A. Lazarian, *ApJ* **614**, 756 (2004).
10. V. L. Ginzburg, S. I. Syrovatskii, *The Origin of Cosmic Rays* (Pergamon Press, Oxford, 1964).
11. *Space Sci. Rev.* **99**, 1-373 (2001).
12. T. A. Lozinskaya, *Supernovae and Stellar Wind in the Interstellar Medium* (AIP, New York, 1992).
13. T. Jones et al., *ApJ* **587**, 227 (2003).
14. J. Vink, *Adv. Space Sci.* **33**, 356 (2004).
15. F. Ahoranian et al., *ApJ* **636**, 777 (2006).
16. F. Ahoranian et al., *A&A* **449**, 223 (2006).
17. H. Katagiri et al., *ApJ* **619**, L163 (2005).
18. F. Ahoranian et al., *A&A* **437**, L7 (2005).
19. J. P. Meyer et al., *ApJ* **487**, 182 (1997).
20. D. C. Ellison et al., *ApJ* **487**, 197 (1997).
21. J. C. Higdon et al., *ApJ* **509**, L33 (1998).
22. J. C. Higdon, R. E. Lingenfelter, *ApJ* **628**, 738 (2005).
23. M. E. Wiedenbeck et al. in *Constraintas on Cosmic-Ray Acceleration and Transport from Isotope Observations*, ed. R. A. Mewaldt et al. (AIP, NY, 2000) p. 363.
24. M. A. Malkov, L. O'C Drury, *Rep. Progr. Phys.* **64**, 429 (2001).
25. J. F. McKenzie, H. J. Völk, *A&A* **116**, 191 (1982).
26. S. G. Lucek, A. R. Bell, *MNRAS* **314**, 65 (2000).

27. A. R. Bell, *MNRAS* **353**, 550 (2004).
28. V. N. Zirakashvili, V. S. Ptuskin, *A&A* (2003).
29. A. Bamba et al., *ApJ* **589**, 253 (2003).
30. H. J. Völk et al., *A&A* **433**, 229 (2005).
31. A. Castellina, *Nucl. Phys. Suppl.* **97**, 35 (2001).
32. A. Haungs et al., *Acta Phys. Polonica B* **35**, 331 (2004).
33. M. Nagano, A. A. Watson, *Rev. Mod. Phys.* **72**, 689 (2000).
34. D. F. Torres, L. A. Anchordoqui, *Rep. Prog. Phys.* **67**, 1663 (2004).
35. D. R. Bergman, *Proc. CRIS 2006, Catania*, in press; preprint astro-ph/0609453 (2006).
36. V. S. Ptuskin, V. N. Zirakashvili, *A&A* **429**, 755 (2005).
37. V. S. Ptuskin et al. *A&A* **268**, 726 (1993).
38. J. R. Hörandel et al., Preprint astro-ph/0609490 (2006).
39. J. R. Hörandel, *Astropart. Phys.* **19**, 193 (2003).
40. A. M. Hillas, *J. Phys. G* **31**, 95 (2005).
41. L. G. Sveshnikova, *A&A* **409**, 799 (2003).
42. A. M. Bykov, I. N. Toptygin, *Astron. Lett.* **27**, 625 (2001).
43. J. R. Jokipii, G. Morfill, *ApJ* bf 290, L1 (1985).
44. V. N. Zirakashvili, H. J. Völk, *Adv. Space Res.* **37**, 1923 (2005).
45. V. S. Berezinsky et al., *Phys. Lett. B* **612**, 147 (2005).
46. E. Khan et al., *Astropart. Phys* **20**, 53 (2005).

COSMIC RAYS FROM THE KNEE TO THE SECOND KNEE: 10^{14} TO 10^{18} eV

JÖRG R. HÖRANDEL

University of Karlsruhe, Institute for Experimental Nuclear Physics, P.O. 3640 76021 Karlsruhe, Germany — www-ik.fzk.de/~joerg

The energies of cosmic rays, fully ionized charged nuclei, extend over a wide range up to 10^{20} eV. A particularly interesting energy region spans from 10^{14} to 10^{18} eV, where the all-particle energy spectrum exhibits two interesting structures, the 'knee' and the 'second knee'. An explanation of these features is thought to be an important step in understanding of the origin of the high-energy particles. Recent results of air shower experiments in this region are discussed. Special attention is drawn to explain the principle of air shower measurements — a simple Heitler model of (hadronic) air showers is developed.

Keywords: cosmic rays, knee, air showers, Heitler model

1. Introduction

The energy spectrum of cosmic rays (fully ionized atomic nuclei) spans a wide range in energy from GeV energies up to 10^{20} eV. Over these 10 decades the flux decreases by about 30 orders of magnitude rather featureless, following roughly a power law $dN/dE \propto E^{\gamma}$. The power law behavior indicates a non-thermal origin of the particles. To reveal small structures in the shape of the energy spectrum the flux is usually multiplied by the energy to some power. The energy spectrum multiplied by E^3 is depicted in Fig. 1. In this representation the spectrum looks rather flat and fine structures can be recognized, indicating small changes in the spectral index γ. The most important are the *knee* at $E_k \approx 4.5$ PeV where the power law spectral index changes from $\gamma = -2.7$ at low energies to $\gamma \approx -3.1$, the *second knee* at $E_{2nd} \approx 400$ PeV$\approx 92 \times E_k$, where the spectrum exhibits a second steepening to $\gamma \approx -3.3$, and the *ankle* at about 4 EeV, above this energy the spectrum seems to flatten again to $\gamma \approx -2.7$. To understand the origin of these structures is expected to be a key element in the understanding of the origin of cosmic rays (CRs).

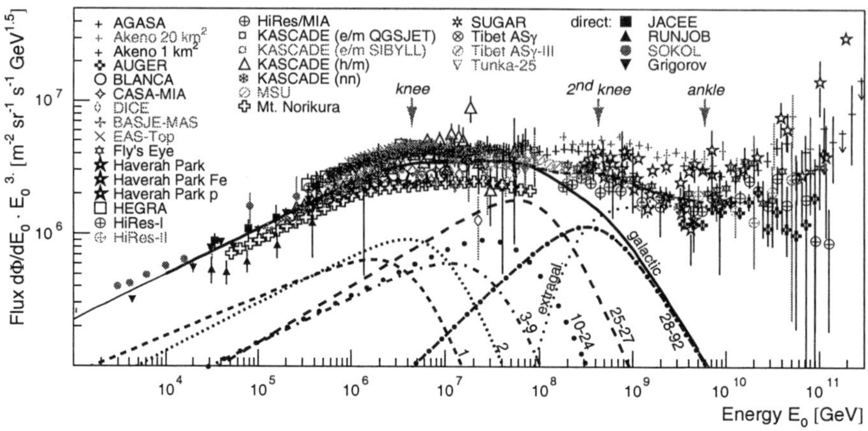

Fig. 1. All-particle energy spectrum of cosmic rays, the flux is multiplied by E^3, for references see [1]. The lines represent spectra for elemental groups (with nuclear charge numbers Z as indicated) according to the poly-gonato model [2]. The sum of all elements (galactic) and a presumably extragalactic component are shown as well.

The lecture starts with a short overview on the physics of galactic cosmic rays (Sect. 2). Measurements in the energy region of interest are performed with air shower experiments, their principles are outlined in Sect. 3 with a simple Heitler model. Finally, recent results of air shower experiments are reviewed in Sect. 4.

2. Galactic cosmic rays and the knee

2.1. *Sources*

At energies around 1 GeV/n all elements known from the periodic table with nuclear charge number Z from 1 to 92 have been found in CRs [2–4]. Overall, the abundance of elements in CRs is very similar to the abundance found in the solar system, which indicates that CRs are "regular matter" but accelerated to very high energies. This is emphasized by measurements of the CRIS experiment [5] which show that the abundances of particular isotopes in cosmic rays and in the solar system differ by less than 20%.

The bulk of CRs is assumed to be accelerated in blast waves of supernova remnants (SNRs). This goes back to an idea by Baade and Zwicky who proposed SNRs as cosmic-rays sources due to energy balance considerations [6]. They realized that the power necessary to sustain the cosmic-ray flux could be provided when a small fraction $\sim 10\%$ of the kinetical energy released in supernova explosions is converted into CRs. Fermi proposed a mechanism to accelerate particles with moving magnetic clouds [7]. This led to todays picture that the particles are accelerated at strong shock fronts in SNRs through first-order Fermi acceleration [8–12]. This theory predicts spectra at the sources following a power law $dN/dE \propto E^{-2.1}$.

Diffusive, first-order shock acceleration works by virtue of the fact that particles gain an amount of energy $\Delta E \propto E$ at each cycle, when a cycle consists of a particle passing from the upstream (unshocked) region to the downstream region and back. At each cycle, there is a probability that the particle is lost downstream and does not return to the shock. Higher energy particles are those that remain longer in the vicinity of the shock and have enough time to achieve the high energy. After a time T the maximum energy attained is $E_{max} \sim Ze\beta_s BTV_s$, where $\beta_s = V_s/c$ is the velocity of the shock. This results in an upper limit, assuming a minimal diffusion length equal to the Larmor radius of a particle of charge Ze in the magnetic fields B behind and ahead of the shock. Using typical values of Type II supernovae exploding in an average interstellar medium yields $E_{max} \approx Z \cdot 100$ TeV [13]. More recent estimates give a maximum energy up to one order of magnitude larger for some types of supernovae $E_{max} \approx Z \cdot 5$ PeV [14–16]. As the maximum energy depends on the charge Z, heavier nuclei (with larger Z) can be accelerated to higher energies. This leads to consecutive cut-offs of the energy spectra for individual elements proportional to their charge Z, starting with the proton component.

This theory is strongly supported by recent measurements of the HESS experiment [17,18], observing TeV γ-rays from the shell type SNR RX J1713.7-3946. For the first time, a SNR could be spatially resolved in γ-rays and spectra have been derived directly at a potential cosmic-ray source. The measurements yield a spectral index $\gamma = -2.19 \pm 0.09 \pm 0.15$ for the observed γ-ray flux. The results are compatible with a nonlinear kinetic theory of cosmic-ray acceleration in supernova remnants and imply that this supernova remnant is an effective source of nuclear CRs, where about 10% of the mechanical explosion energy are converted into nuclear CRs [19].

2.2. *Propagation*

After acceleration, the particles propagate in a diffusive process through the Galaxy, being deflected many times by the randomly oriented magnetic fields ($B \approx 3\ \mu$G). The nuclei are not confined to the galactic disc, they propagate in the galactic halo as well. The diffuse γ-ray background, extending well above the disc, detected by the EGRET experiment, exhibits a structure in the GeV region, which is interpreted as indication for the interaction of propagating CRs with interstellar matter [20]. The γ-rays are produced in inelastic hadronic reactions of CRs with the interstellar medium (ISM) via neutral pion decay $p + \text{ISM} \rightarrow \pi^0 \rightarrow \gamma\gamma$. The height of the propagation region in the halo has been estimated measuring the ^{10}Be/^{9}Be-ratio with the ISOMAX experiment to be a few kpc [21]. Determining the abundance of radioactive nuclei, which decay on the way from the source to the Earth, allows to determine the residence time of CRs in the Galaxy. Measurements with the CRIS instrument yield about $15 \cdot 10^6$ a for particles with GeV energies [22].

Information on the propagation pathlength of CRs is often derived from the measurement of the ratio of primary to secondary nuclei. The latter are produced through spallation during propagation in the Galaxy. The energy dependence of the measured ratio is frequently explained in Leaky Box models by a decreasing pathlength of CRs in the Galaxy $\Lambda(R) = \Lambda_0(R/R_0)^{-\delta}$, with typical values $\Lambda_0 \approx 10 - 15$ g/cm^2, $\delta \approx 0.5 - 0.6$, and the rigidity $R_0 \approx 4$ GV [23]. In this picture the spectra observed at Earth should be steeper as compared to the source, i.e. the spectral index γ should be smaller by the value of δ.

Energy spectra of individual elements have been measured up to energies of about 10^{14} eV by experiments above the atmosphere, the results being well compatible with power laws [2,4]. Due to spallation during the propagation process, the spectra of heavy elements are slightly flatter as compared to light nuclei [2,24], e.g. comparing protons $\gamma_p = -2.71 \pm 0.02$ to iron nuclei $\gamma_{Fe} = -2.59 \pm 0.06$.

The regular component of the galactic magnetic field will cause particles with charge Z to describe helical trajectories with a Larmor radius $R_L = p/(ZeB_0) = 1.08\,\text{pc}\cdot E[\text{PeV}]/(Z\cdot B_0[\mu\text{G}])$, while the random field component causes diffusive propagation. With increasing energy (or momentum) it becomes more difficult to magnetically confine the particles to the Galaxy. Since $R_L \propto 1/Z$ it is expected that leakage from the Galaxy occurs for light elements (low Z) earlier as compared to heavy nuclei, i.e. protons leak first and subsequently all other elements start to escape from the Galaxy.

2.3. *Structures in the energy spectrum*

Many possible origins for the knee are discussed in the literature [25,26]. Most popular are assumptions of a finite energy attained during the acceleration process and leakage from the Galaxy as discussed. In both scenarios the energy spectra of elements exhibit a cut-off at an energy proportional to the nuclear charge Z and the knee in the all-particle spectrum is caused by the cut-off of protons.

All other elements follow subsequently and above a certain energy no more particles are left. On the other hand, the measured all-particle flux extends up to 10^{20} eV, and the highest-energy particles are usually being considered of extragalactic origin. The Larmor radius of a proton with an energy of 10^{20} eV in the galactic magnetic field is $R_L \approx 36$ kpc, comparable to the diameter of the Galaxy. This emphasizes that such high-energy particles are of extragalactic origin. The transition region from galactic to extragalactic CRs is of particular interest, key features are the origin of the second knee and the ankle.

Reviewing the properties of CRs accelerated in SNRs, Hillas finds that a second (galactic) component is necessary to explain the observed flux at energies above 10^{16} eV [27]. Another possibility is a significant contribution of ultra-heavy elements (heavier than iron) to the all-particle flux at energies around 400 PeV [2,24]. In this approach the second knee is caused by the fall-off of the heaviest elements with Z up to 92. It is remarkable that the second knee occurs at $E_{2nd} \approx 92 \times E_k$, the latter being the energy of the first knee. The dip seen in the spectrum between 10^{18} and 10^{19} eV, see Fig. 1, is proposed to be caused by electron-positron pair production of CRs on cosmic microwave background photons [28] $p + \gamma_{3K} \rightarrow p + e^+ + e^-$.

3. Measurement techniques

To clarify the situation and to distinguish between the different models, measurements of the flux of individual elements, or at least groups of elements, up to high energies are necessary. Direct measurements above the atmosphere on stratospheric balloons up to energies exceeding 10^{14} eV are performed with various instruments like ATIC [29], CREAM [30], BESS [31], or TRACER [32]. The presently largest experiment with single-element resolution, TRACER, has an aperture of 5 m^2 sr. With an exposure of 50 m^2 sr d accumulated during a circumpolar flight in 2003 energy spectra could be measured up to about $5 \cdot 10^{14}$ eV for oxygen and $8 \cdot 10^{13}$ eV for iron nuclei [33].

To extend the measurements to energies beyond the knee, at present, ground based installations are the only possibility. With these experiments, secondary products generated in the atmosphere are measured, the extensive air showers (EAS). Air showers were discovered in 1938 by W. Kolhörster [34] and independently by P. Auger [35]. Auger describes his work in a book [36] translated by the director of this school M.M. Shapiro.

In todays experiments, the energy is basically derived from the number of particles observed and the primary's mass is estimated by measurements of the depth of the shower maximum (Sect. 3.1.5) or the electron-to-muon-ratio (Sect. 3.1.6). Two types of experiments may be distinguished: installations measuring the longitudinal development of showers in the atmosphere and apparatus measuring the density (and energy) of secondary particles at ground level.

An example for the latter is the KASCADE experiment [37], covering an area of 200×200 m^2. The basic idea is to measure the electromagnetic component in an array of unshielded scintillation detectors and the muons in scintillation counters shielded by a lead and iron absorber, while the hadronic component is measured in a large calorimeter [38]. The total number of particles at observation level is obtained through the measurement of particle densities and the integration of the lateral density distribution [39]. The direction of air showers is reconstructed through the measurement of the arrival time of the shower particles in the individual detectors.

The depth of the shower maximum is measured in two ways. Light-integrating Čerenkov detectors like the BLANCA [40] or TUNKA [41] experiments are in principle arrays of photomultiplier tubes with light collection cones looking upwards in the night sky, measuring the lateral distribution of Čerenkov light at ground level. The depth of the shower maximum and the shower energy is derived from these observations. Imaging telescopes as in the HiRes [42] or AUGER [43] experiments observe an image of the shower on the sky through measurement of fluorescence light, emitted by nitrogen molecules, which had been excited by air shower particles.

3.1. *A Heitler model for air showers*

The basic properties of EAS are illustrated using a Heitler model [44], expanding an approach by Matthews [45]. The principle ideas of the model are emphasized by full EAS simulations using the CORSIKA code [46] with the hadronic interaction models FLUKA [47] and QGSJET 01 [48]. For the latter, a modification with lower cross-sections has been used [49]. [a]

3.1.1. *Electromagnetic cascades*

A simple approximation of an electromagnetic cascade is shown schematically in Fig. 2. A primary photon generates an e^+e^- pair. An electron radi-

[a]Vertical showers with fixed energies between 10^5 and $3.16 \cdot 10^{10}$ GeV in steps of half a decade have been calculated. Thresholds for photons, electrons, muons, and hadrons were chosen as $E_\gamma > 0.25$ MeV, $E_e > 0.25$ MeV, $E_\mu > 100$ MeV, and $E_h > 100$ MeV.

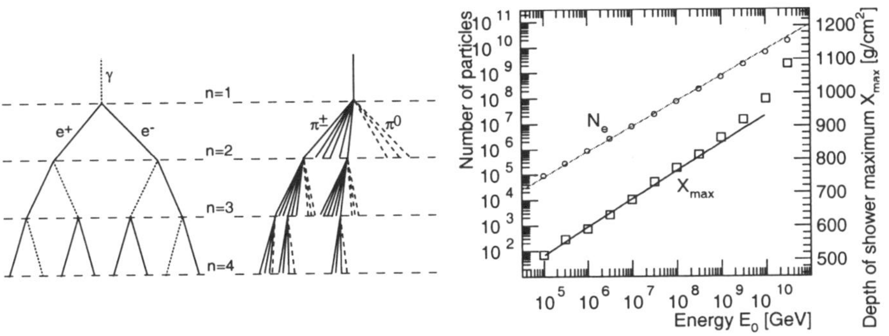

Fig. 2. *Left*: Schematic view of an electromagnetic cascade (*left*) and a hadronic shower (*right*). Not all pion lines are shown [45]. *Right*: Number of electrons at shower maximum and depth of the shower maximum as function of photon energy. The lines are according to (1) and (2).

ates a single photon after traveling one splitting length $d = X_0 \ln 2$, where X_0 is the radiation length ($X_0^{air} = 36.66$ g/cm^2). An electron looses on average half of its energy through radiation over the distance d. After traveling the same distance a photon splits into an e^+e^- pair. In either instance, the energy of a particle is assumed to be equally divided between two outgoing particles. After n splitting lengths, at a distance $x = nX_0 \ln 2$, the total shower size (electrons and photons) is $N = 2^n = \exp(x/X_0)$ and the initial energy E_0 is distributed over N particles. The splitting continues until the energy per particle E_0/N is too low for pair production or bremsstrahlung. Heitler takes this energy to be the critical energy ($E_c^e = 85$ MeV in air), at which ionization losses and radiative losses are equal.

A shower initiated by a primary photon reaches its maximum size N_{max} when all particles have the energy E_c^e, which means $E_0 = E_c^e N_{max}$. The penetration depth X_{max} at which the shower reaches its maximum is obtained by determining the number n_c of splitting lengths, required to reduce the energy per particle to E_c^e. Since $N_{max} = 2^{n_c}$, the number of splitting lengths is $n_c = \ln(E_0/E_c^e)/\ln 2$, giving $N_{max} = E_0/E_c^e$ and

$$X_{max}^{\gamma} = n_c X_0 \ln 2 = X_0 \ln(E_0/E_c^e). \tag{1}$$

The elongation rate Λ specifies the increase of X_{max} with energy E_0 and is defined as $\Lambda \equiv \mathrm{d}X_{max}/\mathrm{d}\,\mathrm{lg}E_0$. Using (1) gives $\Lambda^{\gamma} = \ln 10\, X_0 = 84.4$ g/cm^2 per decade of primary energy for electromagnetic showers in air. Thus, $X_{max} = 597$ g/cm$^2 + 84$ g/cm$^2 \,\mathrm{lg}(E_0/\mathrm{PeV})$ is expected. This prediction agrees well with results of full simulations as can be inferred from Fig. 2. [b]

The simple model describes quite well the position of the maximum of electromagnetic cascades when compared to EAS simulations and to measurements at accelerators [45,52]. However, the model overestimates the actual ratio of electrons to photons. It predicts that after a few generations the electron size approaches $N_e \approx \frac{2}{3}N_{max}$. This is much too large for several reasons, mainly that multiple photons are often radiated during bremsstrahlung and many electrons and positrons range out in the air.

To extract the number of electrons N_e^{max} at shower maximum from Heitler's total size N_{max}, a simple correction

$$N_e^{max} = \frac{E_0}{gE_c^e} \tag{2}$$

is adopted, with a constant value g. When the estimated electron number is compared to measurements, the factor g has to be fine tuned. It depends on properties of the detectors used like the energy threshold and the efficiency to detect photons and electrons (or positrons). Comparisons with results at accelerators indicate values between $g = 10$ [45] and $g = 20$ [52]. Results of a full EAS simulation are depicted in Fig. 2. For electromagnetic showers the number of electrons turns out to be almost exactly linearly proportional to the shower energy as expected from (2). A fit yields $N_e \propto E_0^{0.97}$ and a correction factor $g \approx 13$ is obtained, compatible with the accelerator based results. With this value the number of electrons at shower maximum is according to (2) $N_e^{max} = 9.0 \cdot 10^5 \cdot E_0/\text{PeV}$.

3.1.2. *Hadronic showers*

Hadron induced showers are modeled using a similar approach, for a figurative sketch, see Fig. 2. The atmosphere is divided in layers of fixed thickness $\lambda_i \ln 2$, where λ_i is the interaction length of strongly interacting particles. An energy around 100 GeV is a typical energy for pions in air showers and for a simple approach a constant value $\lambda_i = 120$ g/cm^2 is adopted. Hadrons interact after traversing one layer, producing N_{ch} charged pions and $\frac{1}{2}N_{ch}$ neutral pions. The latter decay promptly to photons, initiating electromagnetic cascades. Charged pions continue through another layer and interact. The process continues until the charged pions fall bellow the critical energy E_c^π, where they are all assumed to decay, yielding muons.

[b]The deviations at high energies are due to the Landau Pomeranchuk Migdal effect [50,51].

The multiplicity of charged particles produced in hadron interactions increases very slowly with laboratory energy $\propto E^{0.2}$ in pp and $p\bar{p}$ data [53]. The multiplicity in π-^{14}N collisions increases as $N_{ch} \approx 5$, 11, and 27 at 10, 100, and 10^4 GeV, respectively [49]. A constant value $N_{ch} = 10$ is adopted in the following for the number of charged particles produced in pion-air interactions, again corresponding to an energy of about 100 GeV.

The second parameter is the energy E_c^π at which further particle production by $\pi^\pm$ ceases. E_c^π may be defined as the energy at which the probability for decay and hadronic interaction equalize. Following Ref. [45] a constant critical pion energy $E_c^\pi = 20$ GeV is adopted in the following.

If we consider a proton with E_0 entering the atmosphere, we have after n interactions $N_\pi = (N_{ch})^n$ charged pions. Assuming equal division of energy during particle production, these pions carry a total energy of $(2/3)^n E_0$. The remainder of the energy goes into electromagnetic showers from π^0 decays. Hence, the energy per charged pion is $E_\pi = E_0/(\frac{3}{2}N_{ch})^n$. After a certain number n_c of generations, E_π becomes less than E_c^π. The number of interactions needed to reach $E_\pi = E_c^\pi$ is

$$n_c = \frac{\ln E_0/E_c^\pi}{\ln \frac{3}{2} N_{ch}} = 0.85 \lg \left(\frac{E_0}{E_c^\pi} \right). \tag{3}$$

3.1.3. *Number of muons*

The number of muons is obtained, assuming that all pions decay, using $N_\mu = N_\pi = (N_{ch})^{n_c}$. Their energy dependence is derived applying (3)

$$\ln N_\mu = n_c \ln N_{ch} = \beta \ln \left(\frac{E_0}{E_c^\pi} \right), \quad \text{with} \quad \beta = \frac{\ln N_{ch}}{\ln \frac{3}{2} N_{ch}} \approx 0.85 \tag{4}$$

for $N_{ch} = 10$. It should be noted that although N_{ch} changes (slowly) as the shower develops, β depends only logarithmically on this value.

So far, an important aspect of hadronic interactions has been neglected. In an interaction only a fraction of the energy is available for secondary particle production, usually characterized by the the inelasticity κ. Taking this effect into account, in an interaction initiated by a particle with energy E, the energy $(1-\kappa)E$ is taken away by a single leading particle, $\frac{2}{3}\kappa E$ is used to produce N_{ch} charged pions, and $\frac{1}{3}\kappa E$ goes via neutral pions into the electromagnetic component. Including inelasticity in the Heitler model changes the parameter β in (4) to [45]

$$\beta = \frac{\ln[1 + N_{ch}]}{\ln\left[(1 + N_{ch})/(1 - \frac{1}{3}\kappa)\right]} \approx 1 - \frac{\kappa}{3\ln(N_{ch})} = 1 - 0.14\kappa. \tag{5}$$

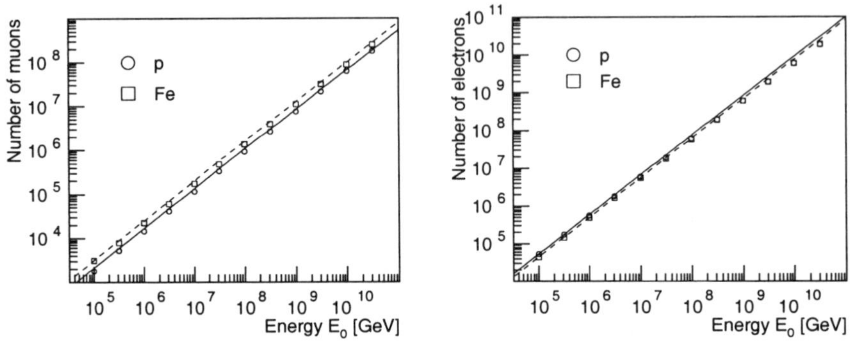

Fig. 3. Number of muons (*left*) and number of electrons (*right*) at shower maximum as function of energy for primary protons and iron nuclei according to CORSIKA simulations (symbols). The lines are predictions according to (6) and (10), respectively, for protons (—) and iron (- - -) nuclei.

The elasticity for the most energetic meson in pion-air interactions yields $1-\kappa$ between 0.26 and 0.32 [49], resulting in $\beta = 0.90$.

To expand the simple approach from primary protons to nuclei, the superposition model is used. A nucleus with atomic mass number A and energy E_0 is taken to be A individual single nucleons, each with energy E_0/A, and each acting independently. The resulting EAS is treated as the sum of A individual proton induced showers, all starting at the same point. The observable shower features are obtained by substituting the lower primary energy into the expressions derived for proton showers and summing A such showers. Applying this to the number of muons yields $N_\mu = A(E_0/(AE_c^\pi))^\beta$. The number of muons in showers induced by nuclei with mass number A and energy E_0 is then

$$N_\mu = \left(\frac{E_0}{E_c^\pi}\right)^\beta A^{1-\beta} \approx 1.69 \cdot 10^4 \cdot A^{0.10} \left(\frac{E_0}{1\ \mathrm{PeV}}\right)^{0.90} . \tag{6}$$

Two important features follow from (6): the number of muons increases as function of energy slightly less than exactly linear and N_μ increases as function of the mass of the primary particle as $\propto A^{0.1}$. Accordingly, iron induced showers contain about 1.5 times as many muons as proton showers with the same energy. This results from the less-than-linear growth of the number of muons with energy – $\beta < 1$ in (6). The lower energy nucleons which initiate the shower generate fewer interaction generations, and consequently, loose less energy to the electromagnetic component.

The number of muons at shower maximum as function of energy is shown in Fig. 3 as obtained from full simulations. The lines indicate predictions according to (6), being well in agreement with the simulations.

3.1.4. *Number of electrons*

Conservation of energy implies that the primary energy is split into electromagnetic and hadronic parts $E_0 = E_{em} + E_h$. The number of electrons is estimated using this relation. The hadronic energy appears in the simple approach in the muon component as $E_h = N_\mu E_c^\pi$ and the energy fraction for the electromagnetic component is, using (6)

$$\frac{E_{em}}{E_0} = \frac{E_0 - N_\mu E_c^\pi}{E_0} = 1 - \left(\frac{E_0}{A\,E_c^\pi}\right)^{\beta-1} . \tag{7}$$

The electromagnetic fraction is 57% at $E_0 = 10^{14}$ eV, increasing to 79% at 10^{17} eV for proton induced showers. For iron induced showers the fraction rises from 38% to 68%.

Equation (7) can be approximated by a power law

$$\frac{E_{em}}{E_0} \approx a\left(\frac{E_0}{A\,E_c^\pi}\right)^{b} . \tag{8}$$

Series expansion near $x_0 = E_0/E_c^\pi = 10^5$ yields the number of electrons at shower maximum as function of energy

$$N_e = \frac{E_{em}}{gE_c^e} \approx \frac{a}{gE_c^e}\left(A\,E_c^\pi\right)^{-b} E_0^{1+b} \tag{9}$$

with $b = (1-\beta)/(x_0^{1-\beta}-1) \approx 0.046$ and $a = (1-x_0^{\beta-1})/(x_0^b) \approx 0.40$. Hence, $\alpha = 1 + b \approx 1.046$ is obtained, which leads, using $g = 13$ to

$$N_e \approx 5.95 \cdot 10^5 \cdot A^{-0.046}\left(\frac{E_0}{1\ \text{PeV}}\right)^{1.046} . \tag{10}$$

This implies that the number of electrons grows as function of energy slightly faster than exactly linear. The electron number decreases with increasing mass number, an iron induced shower is expected to contain about 83% of the electromagnetic energy of a proton shower with the same energy. It should be emphasized that the model does not take into account absorption in the atmosphere, thus, the number of electrons obtained is valid at shower maximum.

The number of electrons at shower maximum according to full simulations is shown as function of energy in Fig. 3. The results are compared to predictions according to (10) for proton and iron induced showers indicated by the lines. It can be seen that the simple model reproduces quite well the results of the full simulations.

3.1.5. *Depth of the shower maximum*

The atmospheric depth at which the electromagnetic shower component reaches its maximum is called X_{max}. In hadronic interactions $\kappa/3$ of the available energy goes into the electromagnetic component via π^0-decays, see Fig. 2. For a simple estimate only the first generation of electromagnetic showers is used. This approach will certainly underestimate the value of X_{max} since it neglects the following subshowers.

The first interaction occurs at an atmospheric depth $X_1 = \lambda_i \ln 2$, where λ_i is the interaction length of a primary proton λ_i^{p-air}. The latter can be approximated around 1 PeV by the relation

$$\lambda_i^{p-air} = \xi + \zeta \lg \frac{E_0}{1\ \text{PeV}} \tag{11}$$

with $\xi = 68.55$ g/cm^2 and $\zeta = -4.88$ g/cm^2.

In the first interaction $\frac{1}{2}N_{ch}$ neutral pions are produced, yielding N_{ch} photons. Each photon initiates an electromagnetic cascade with the energy $\kappa E_0/(3N_{ch})$, developing in parallel with the others. The average multiplicity of charged particles produced in pion-nitrogen interactions [49] can be parameterized for energies around 1 PeV as

$$N_{ch} = N_0 \left(\frac{E_0}{1\ \text{PeV}}\right)^{\eta} \tag{12}$$

with $N_0 = 55.2$ and $\eta = 0.13$.

The depth of the shower maximum is obtained as in (1) for an electromagnetic shower with an energy $\kappa E_0/(3N_{ch})$, starting after the first interaction at a depth X_1, $X_{max}^p = \lambda_i^{p-air} \ln 2 + X_0 \ln(\kappa E_0/(3N_{ch}E_c^e))$. Using (11) and (12), the expression

$$X_{max}^p = \xi \ln 2 - X_0 \ln\left(\frac{3N_0}{\kappa}\frac{E_c^e}{\text{PeV}}\right) + (X_0 \ln 10 - \eta X_0 \ln 10 + \zeta \ln 2) \lg \frac{E_0}{\text{PeV}} \tag{13}$$

is obtained. The elongation rate for protons is determined by the elongation rate for electromagnetic showers $\Lambda^\gamma = X_0 \ln 10$ and in addition by terms which take into account the growing multiplicity of secondary particles, as well as the decreasing interaction length as function of energy. Taking the numerical parameters as described, the elongation rate $\Lambda^p = (84.4 - 11.0 - 3.4)$ g/cm^2 $= 70.0$ g/cm^2 is obtained, and one realizes that the effect of growing multiplicity dominates the effect of a decreasing interaction length by about a factor three. Evaluating also the constant term in (13) yields $X_{max}^p = 442.9$ g/cm$^2 + 70.0$ g/cm$^2 \lg(E_0/\text{PeV})$.

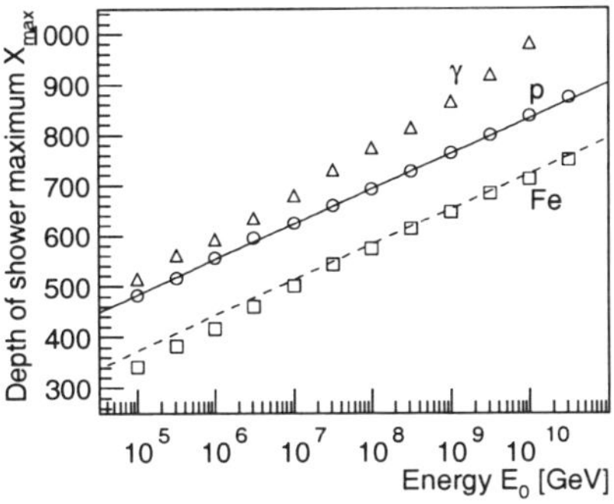

Fig. 4. Average depth of the shower maximum for primary photons, protons, and iron nuclei according to CORSIKA simulations. The lines indicate predictions according to (13) (- - -) and the same function shifted up by 110 g/cm^2 (—).

When compared to results of full simulations, the calculated values for X_{max}^p are about 110 g/cm^2 or almost $2\lambda_i^{p-air}$ smaller than the results of full calculations depicted in Fig. 4 (dashed line). Presumably this is a consequence of neglecting the contributions of following generations of π^0 production. However, the predicted elongation rate agrees extremely well with the value obtained from the CORSIKA simulations at 1 PeV $\Lambda^p = 69.9 \pm 0.3$ g/cm^2 per decade. The solid line represents (13) shifted upwards by 110 g/cm^2 and agrees well with the proton simulations.

To expand the simple approach from primary protons to nuclei with mass number A, the superposition model is used and in (13) the energy E_0 is substituted by E_0/A. This yields $X_{max}^A = X_{max}^p - X_0 \ln A$, predicting that the maximum for iron induced showers should be about 150 g/cm^2 higher up in the atmosphere. In the full simulations, the difference is slightly smaller as can be inferred from the figure.

3.1.6. *Energy and mass of the primary particle*

In EAS experiments the reconstructed number of electrons and muons are often presented in the lgN_μ-lgN_e plane in order to estimate the energy and mass of the shower inducing particles. As an application of the simple Heitler Model, lines of constant mass and energy in the N_μ-N_e plane are derived in the following to illustrate the method utilized in the experiments.

To deduce lines of constant mass, (9) is transformed to obtain E_0, which, in turn is introduced into (6). This yields the number of muons as function of the number of electrons at shower maximum

$$N_\mu|_A = (E_c^\pi)^{-\delta} \left(\frac{gE_c^e}{a}\right)^\delta A^{1-\delta} N_e^\delta \approx 0.18 \cdot A^{0.14} N_e^{0.86} \qquad (14)$$

with the exponent $\delta = \beta/(1+b) \approx 0.86$. In a similar way, lines of constant

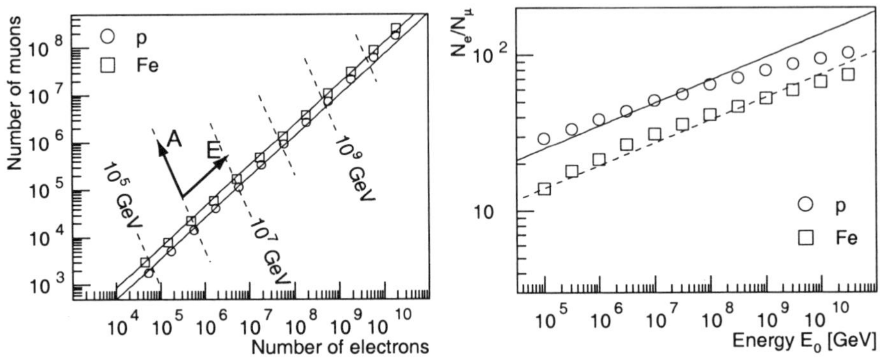

Fig. 5. *Left*: Average number of muons versus number of electrons at shower maximum for primary protons and iron nuclei. The data points are results of full simulations. The solid lines represent (14) for protons and iron nuclei, the dashed lines are equal energy lines according to (15). *Right*: Ratio of electron to muon number N_e/N_μ at shower maximum as function of energy for primary protons and iron nuclei. The data points are results of full simulations. The lines indicate (16) for the two primaries.

energy are derived. A is taken from (9) and put in (6), which leads to

$$N_\mu|_{E_0} = \frac{1}{E_c^\pi}\left(\frac{gE_c^e}{a}\right)^\varepsilon E_0^{\beta+\varepsilon(b-1)} N_e^\varepsilon \approx 5.77 \cdot 10^{16}\left(\frac{E_0}{1\ \mathrm{PeV}}\right)^{2.97} N_e^{-2.17} \tag{15}$$

with an exponent $\varepsilon = -(1-\beta)/b \approx -2.17$.

The constant-mass lines for protons and iron nuclei are shown in Fig. 5 together with equal-energy lines for energies from 10^5 to 10^{10} GeV. These sets of lines form a coordinate system for energy and mass in the N_μ-N_e plane. The axis are non-perpendicular to each other. In the figure also results of full CORSIKA simulations for proton and iron induced showers are shown for fixed energies from 10^5 to $3.16 \cdot 10^{10}$ GeV in steps of half a decade. Taking the simplicity of the model into account the predicted lines agree quite well with the full simulations and they give a good illustration of the physics in the N_μ-N_e plane.

Dividing (9) by (6) yields the electron-to-muon ratio at shower maximum

$$\frac{N_e}{N_\mu} = \frac{a}{gE_c^e}(E_c^\pi)^{\beta-b}\left(\frac{E_0}{A}\right)^{1+b-\beta} \approx 35.1 \cdot \left(\frac{E_0}{1\ \mathrm{PeV} \cdot A}\right)^{0.15} . \tag{16}$$

It depends on the energy per nucleon E_0/A of the primary particle. This is the reason why the ratio N_e/N_μ is frequently used in EAS experiments to estimate the mass of the primary particle. If the energy is derived from another observable, the mass can be inferred. Predictions according to (16) are compared to results of full simulations for proton and iron induced showers in Fig. 5. The simple model predicts the calculated ratio quite well. The CORSIKA simulations exhibit almost a power law behavior, however, at high energies some flattening with respect to the predicted slope is visible.

4. Experimental results

The all-particle energy spectra obtained by many experiments are compiled in Fig. 1. Shown are results from direct measurements above the atmosphere as well as from various air shower experiments. The individual measurements agree within a factor of two or three in the flux and a similar shape can be recognized for all experiments with a knee at energies of about 4 PeV. Typical values for the systematic uncertainties of the absolute energy scale for air shower experiments are about 15 to 20%. Renormalizing the energy scales of the individual experiments to match the all-particle spectrum obtained by direct measurements in the energy region up to almost a PeV requires correction factors of the order of $\pm 10\%$ [2]. Indicating that the all-particle spectrum seems to be well determined.

Due to the large fluctuations in air showers it is not possible to derive energy spectra for individual elements from air shower data. Therefore, frequently the mean mass of CRs is investigated. An often-used quantity to characterize the composition is the mean logarithmic mass, defined as $\langle \ln A \rangle = \sum_i r_i \ln A_i$, r_i being the relative fraction of nuclei of mass A_i. Investigating the ratio of the number of electrons and muons at ground level and the average depth of the shower maximum an increase of the mean logarithmic mass in the energy range around the knee could be observed by many experiments [1]. Such an increase is expected from consecutive cut-offs of the energy spectra of individual elements, starting with protons.

A significant step forward in understanding the origin of CRs are measurements of energy spectra for individual elements or at least groups of elements. Up to about a PeV direct measurements have been performed with instruments above the atmosphere. As examples, results for primary protons, helium, and iron nuclei are compiled in Fig. 6. Recently, also indirect measurements of elemental groups became possible.

With the KASCADE experiment, the problem of composition has been approached in various ways [55]. An advanced analysis is founded on the measurement of the electromagnetic and muonic shower components [54]. It is based on the deconvolution of a two-dimensional electron muon number distribution. Unfolding is performed using two hadronic interaction models

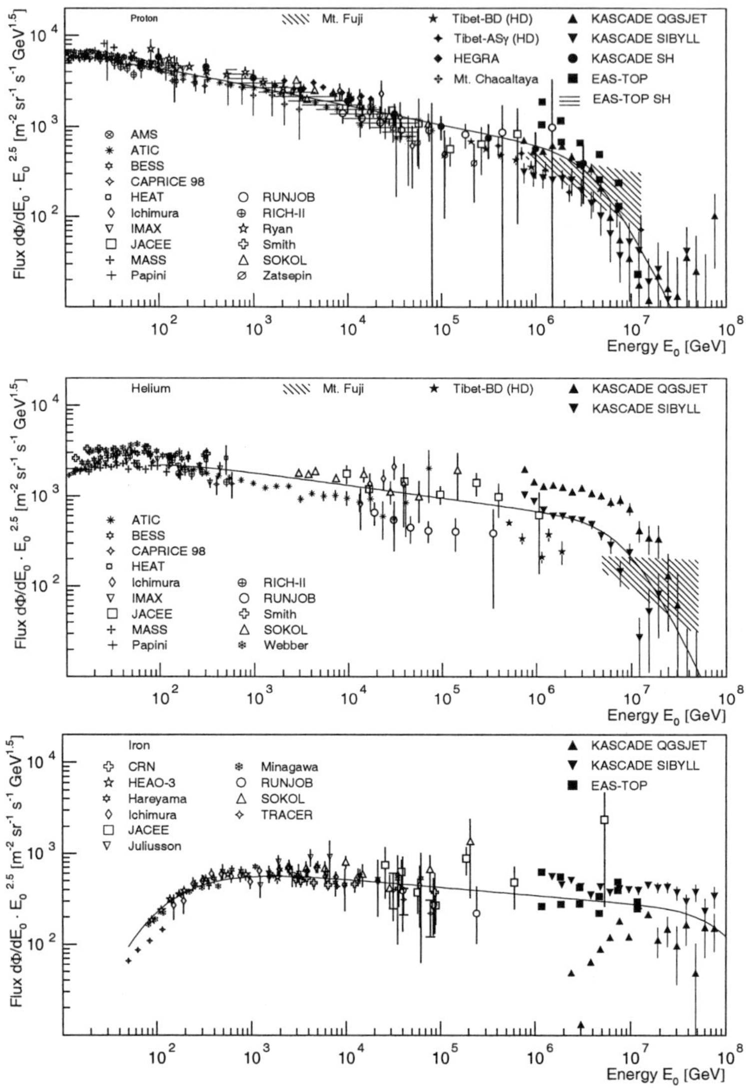

Fig. 6. Energy spectra for primary protons, helium, and iron nuclei from direct and indirect measurements for references see [1]. The lines indicate spectra according to the poly-gonato model [2].

(QGSJET 01 and SIBYLL 2.1) to interpret the data. The spectra obtained for five elemental groups are displayed in Fig. 7. They exhibit sequential cut-offs in the flux for the light elements. For both models a depression

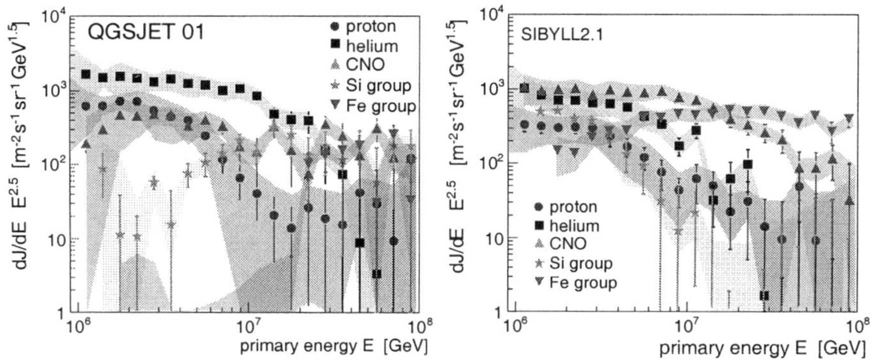

Fig. 7. Energy spectra for elemental groups as obtained by the KASCADE experiment, using two different models (QGSJET 01 and SIBYLL 2.1) to interpret hadronic interactions in the atmosphere [54].

is visible for protons around 3 to 4 PeV and at higher energies for helium nuclei. The systematic differences in flux for the spectra derived with QGSJET and SIBYLL amount to a factor of about two to three. The silicon and iron groups show a rather unexpected behavior for both models. The increase of the flux for both groups (QGSJET) and the early cut-off for the silicon group (SIBYLL) is not compatible with contemporary astrophysical models. The discrepancies are attributed to the fact that none of the air shower models is able to describe the observed data set in the whole energy range consistently [54].

The KASCADE results are compared to results of other experiments in Fig. 6. EAS-TOP derived spectra from the simultaneous observation of the electromagnetic and muonic components. HEGRA used an imaging Čerenkov telescope system to derive the primary proton flux [56]. The primary proton flux has been derived from measurements of the flux of unaccompanied hadrons at ground level with the EAS-TOP and KASCADE experiments [57,58]. Spectra for protons and helium nuclei are obtained from emulsion chambers exposed at Mts. Fuji and Kanbala [59]. The Tibet group performs measurements with a burst detector as well as with emulsion chambers and an air shower array [60,61].

Considering the energy range above 10 GeV, at least a qualitative picture of the energy spectra for individual mass groups emerges: the spectra seem to be compatible with power laws with a cut-off at high energies. The spectra according to the poly gonato model [2] are indicated in the figures as lines. It can be recognized that the measured values are compatible with cut-offs at energies proportional to the nuclear charge $\hat{E}_Z = Z \cdot 4.5$ PeV.

The lines in Fig. 1 indicate spectra for the same model. Summing up the flux of all elements, the all-particle flux is compatible with the flux derived from air shower experiments in the knee region. Above 10^8 GeV the flux of galactic CRs is not sufficient to account for the observed all-particle spectrum, and an additional, presumably extragalactic component is required.

Energy spectra have been reconstructed with KASCADE data up to energies of 100 PeV. At these energies statistical errors start to dominate the overall error. To improve this situation, the experiment has been enlarged. Covering an area of 0.5 km^2, 37 detector stations, containing 10 m^2 of plastic scintillators each, have been installed to extend the original KASCADE set-up [62]. Regular measurements with this new array and the original KASCADE detectors, forming the KASCADE-Grande experiment, are performed since summer 2003 [63]. The objective is to reconstruct energy spectra for groups of elements up to 10^{18} eV [64], covering the energy region of the second knee, where the galactic cosmic ray spectrum is expected to end [65]. First analyses extend the lateral distributions of electrons and muons up to 600 m [66,67]. Based on one year of measurements, already energies close to 10^{18} eV are reached. It is planned to conduct an unfolding analysis, similar to the one described above, and reveal the energy spectra for groups of elements up to 10^{18} eV.

A more detailed discussion of experimental results may be found elsewhere [1,26,68,69].

5. Conclusion and Outlook

In the last decade the understanding of the origin of high-energy CRs has advanced significantly. In particular, the KASCADE experiment has shown that the origin of the knee in the all-particle energy spectrum is due to a cut-off of the light elements. A corresponding increase of the mean mass as function of energy in the knee region is observed by many experiments. Such a behavior is expected from astrophysical models, explaining the knee due to a finite energy reached in the acceleration process and due to leakage from the Galaxy. However, it has also evolved that the astrophysical interpretation of air shower data, at present, is limited by the understanding of high-energy hadronic interactions in the atmosphere. Experiments like KASCADE have reached the sensitivity to improve interaction models and corresponding analyses are under way, e.g. [70–72]. A big step forward is the observation of TeV-γ-rays from SNRs with the expected spectral index $\gamma \approx -2.1$, thus giving an important hint to the sources of hadronic CRs.

In the next years the KASCADE-Grande experiment and the Ice

Cube/Ice Top experiment at the south pole [73] will measure CRs in the energy region of the second knee and will provide information on the mass composition in this region, where the galactic CR component is expected to end. Balloon borne experiments like ATIC, CREAM, or TRACER will improve the knowledge about CR propagation, by extending the energy spectra of individual elements to energies approaching the knee.

Acknowledgment: It was a pleasure to participate in an inspiring school and to experience the great hospitality in Erice. I would like to thank Jim Matthews for our exchange about the Heitler model and acknowledge the stimulating scientific discussions with my colleagues from the KASCADE-Grande, TRACER, and AUGER experiments.

References

1. J. Hörandel, J. Phys.: Conf. Ser. 47 (2006) 41.
2. J. Hörandel, Astropart. Phys. 19 (2003) 193.
3. M. Shapiro, R. Silberberg, Ann. Rev. Nucl. Sci. 20 (1970) 323.
4. B. Wiebel-Soth et al., Astron. & Astroph. 330 (1998) 389.
5. M. Wiedenbeck et al., 28th Int. Cosmic Ray Conf., Tsukuba 4 (2003) 1899.
6. W. Baade, F. Zwicky, Phys. Rev. 46 (1934) 76.
7. E. Fermi, Phys. Rev. 75 (1949) 1169.
8. W. Axford et al., 15th Int. Cosmic Ray Conf., Plovdiv 11 (1977) 132.
9. G. Krymsky, Dok. Acad. Nauk USSR 234 (1977) 1306.
10. A. Bell, Mon. Not. R. Astr. Soc. 182 (1978) 147.
11. R. Blanford, J. Ostriker, Astrophys. J. 221 (1978) L29.
12. M. Longair, High Energy Astrophysics vol. 2, Cambridge, 1994.
13. P. Lagage, C. Cesarsky, Astron. & Astroph. 125 (1983) 249.
14. E. Berezhko, Astropart. Phys. 5 (1996) 367.
15. K. Kobayakawa et al., Phys. Rev. D 66 (2002) 083004.
16. L. Sveshnikova et al., Astron. & Astroph. 409 (2003) 799.
17. F. Aharonian et al., Nature 432 (2004) 75.
18. F. Aharonian et al., Astron. & Astroph. 449 (2006) 223.
19. H. Völk, E. Berezhko, Astron. & Astroph. 451 (2006) 981.
20. A. Strong, I. Moskalenko, preprint astro-ph/9903370 (1999).
21. A. Molnar, M. Simon, 28th Int. Cosmic Ray Conf., Tsukuba 4 (2003) 1937.
22. N. Yanasak et al., Astrophys. J. 563 (2001) 768.
23. S. Stephens, R. Streitmatter, Astrophys. J. 505 (1998) 266.
24. J. Hörandel et al., astro-ph/0609490 (2006).
25. J. Hörandel, Astropart. Phys. 21 (2004) 241.
26. J. Hörandel, Int. J. Mod. Phys. A 20 (2005) 6753.
27. A. Hillas, J. Phys. G: Nucl. Part. Phys. 31 (2005) R95.
28. V. Berezinsky, astro-ph/0509069 (2005).
29. T. Guzik et al., Adv. Space Res. (2004) in press.
30. E. Seo et al., Adv. Space Res. 33 (2004) 1777.

31. Y. Ajima et al., Nucl. Instr. & Meth. A 443 (2000) 71.
32. F. Gahbauer et al., Astrophys. J. 607 (2004) 333.
33. D. Müller et al., 29th Int. Cosmic Ray Conf., Pune 3 (2005) 89.
34. W. Kolhörster et al., Naturw. 26 (1938) 576.
35. P. Auger et al., Comptes renduz 206 (1938) 1721.
36. P. Auger, What are Cosmic Rays, University of Chicago Press, 1945.
37. T. Antoni et al., Nucl. Instr. & Meth. A 513 (2003) 490.
38. J. Engler et al., Nucl. Instr. & Meth. A 427 (1999) 528.
39. T. Antoni et al., Astropart. Phys. 14 (2001) 245.
40. J. Fowler et al., Astropart. Phys. 15 (2001) 49.
41. O. Gress et al., Nucl. Phys. B (Proc. Suppl.) 75A (1999) 299.
42. T. Abu-Zayyad et al., Astrophys. J. 557 (2000) 686.
43. J. Abraham et al., Nucl. Instr. & Meth. A 523 (2004) 50.
44. W. Heitler, The quantum theory of radiation, Oxford, 1954.
45. J. Matthews, Astropart. Phys. 22 (2005) 387.
46. D. Heck et al., Report FZKA 6019, Forschungszentrum Karlsruhe (1998).
47. A. Fassò et al., FLUKA: Status and Prospective of Hadronic Applications, Proc. Monte Carlo 2000 Conf., Lisbon, A. Kling, F. Barao, M. Nakagawa, P. Vaz eds., Springer (Berlin), 2001, p. 955.
48. N. Kalmykov et al., Nucl. Phys. B (Proc. Suppl.) 52B (1997) 17.
49. J. Hörandel, J. Phys. G: Nucl. Part. Phys. 29 (2003) 2439.
50. L. Landau, I. Pomeranchuk, Dokl. Akad. Nauk SSSR 92 (1953) 535 and 735.
51. A. Migdal, Phys. Rev. 103 (1956) 1811.
52. S. Plewnia et al., Nucl. Instr. & Meth. A 566 (2006) 422.
53. S. Eidelman et al., Phys. Lett. B 592 (2004) 1.
54. T. Antoni et al., Astropart. Phys. 24 (2005) 1.
55. J. Hörandel et al., J. Phys.: Conf. Ser. 39 (2006) 463.
56. F. Aharonian et al., Phys. Rev. D 59 (1999) 092003.
57. M. Aglietta et al., Astropart. Phys. 19 (2003) 329.
58. T. Antoni et al., Astrophys. J. 612 (2004) 914.
59. J. Huang et al., Astropart. Phys. 18 (2003) 637.
60. M. Amenomori et al., Phys. Rev. D 62 (2000) 112002.
61. M. Amenomori et al., 28th Int. Cosmic Ray Conf., Tsukuba 1 (2003) 143.
62. G. Navarra et al., Nucl. Instr. & Meth. A 518 (2004) 207.
63. A. Chiavassa et al., 29th Int. Cosmic Ray Conf., Pune 6 (2005) 313.
64. A. Haungs et al., J. Phys.: Conf. Ser. 47 (2006) 238.
65. J. Hörandel et al., J. Phys.: Conf. Ser. 47 (2006) 132.
66. R. Glasstetter et al., 29th Int. Cosmic Ray Conf., Pune 6 (2005) 293.
67. J. v. Buren et al., 29th Int. Cosmic Ray Conf., Pune 6 (2005) 301.
68. S. Swordy et al., Astropart. Phys. 18 (2002) 129.
69. J. Hörandel, Neutrinos and Explosive Events in the Universe, M.M. Shapiro, T. Stanev, J.P. Wefel (eds.) NATO Science Series, Springer, 2005, p. 365.
70. J. Hörandel et al., Nucl. Phys. B (Proc. Suppl.) 151 (2006) 205.
71. J. Hörandel et al., 29th Int. Cosmic Ray Conf., Pune 6 (2005) 121.
72. J. Milke et al., 29th Int. Cosmic Ray Conf., Pune 6 (2005) 125.
73. T. Gaisser et al., 28th Int. Cosmic Ray Conf., Tsukuba 2 (2003) 1117.

ULTRA HIGH ENERGY COSMIC RAYS: ORIGIN AND PROPAGATION

TODOR STANEV

Bartol Research Institute, Department of Physics and Astronomy, University of Delaware, Newark, DE 19716, U.S.A.

We briefly discuss ideas related to the origin of the ultrahigh energy cosmic rays. If these extremely high energy particles are generated all around the Universe their production spectrum is modified by interactions in the universal photon fields in their propagation from the sources to us. Cosmogenic neutrinos that are generated in these interactions can help us understand better their origin.

1. Introduction

When the first article about the detection of an ultrahigh energy cosmic ray (UHECR) shower appeared in 1963 [1] it did not surprise anybody. Physicists were convinced that the cosmic ray spectrum continues forever and only the low flux is the reason such showers were not detected. Seven years earlier Cocconi published and article [2] that discussed the extragalactic origin of such high energy cosmic rays. This event became much more exciting after Greisen [3] and Zatsepin&Kuzmin [4] published simultaneously papers discussing the propagation of ultra high energy particles in extragalactic space. They calculated the energy loss distance of nucleons interacting in the newly discovered microwave background and reached the conclusion that it is shorter than the distances between powerful galaxies. The cosmic ray spectrum should thus have an end around energy of 5×10^{19} eV. This effect is now known as the GZK cutoff.

The flux of UHECR of energy above 10^{20} eV is estimated to 0.5 to 1 event per square kilometer per century per steradian. Even big detectors of area tens of km^2 would only detect a couple of events for ten years. The topic became one of common interest during the last decade of the last century when ideas appeared for construction of detectors with effective areas in thousand of km^2. Such detectors would detect tens of events per year and finally solve all mysteries surrounding UHECR [5–7].

Cosmic rays come on a featureless, power law like, $F(E) = K \times E^{-\alpha}$, spectrum. There are only two distinct features: the steepening of the spectral index α from 2.7 to 3.1 above 10^6 GeV (*the knee*) and the flattening at about 10^9 GeV.

We believe that cosmic rays below the knee are accelerated at galactic supernovae remnants. Cosmic rays above the knee are also thought to be of galactic origin, although there is no clue of their acceleration sites. Cosmic rays above the ankle are thought to be extragalactic. The transition from galactic to extragalactic cosmic rays is now a topic of high astrophysical interest since it fully defines the extragalactic cosmic ray flux that is related to many important astrophysical parameters.

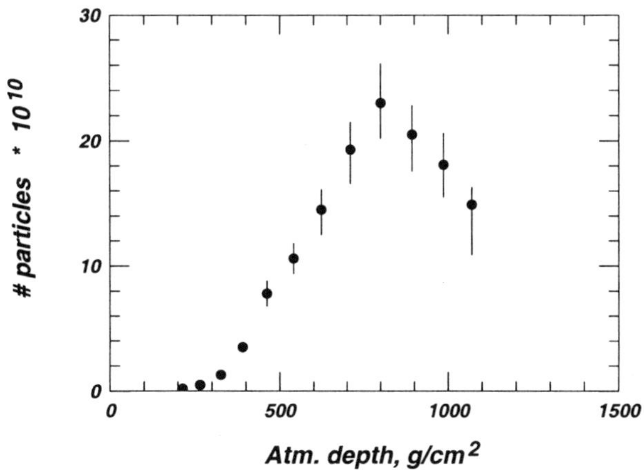

Fig. 1. Shower profile of the highest energy cosmic ray shower detected by the Fly's Eye.

Part of these parameters are related to the cosmic ray acceleration to 10^{20} eV (10^{11} GeV) eleven orders of magnitude higher than the proton mass. Because of the difficulties in modeling such a process there are many non-acceleration models for the production of UHECR. Others are relevant for the understanding of cosmic ray propagation in extragalactic space and the exact results of the interactions in propagation.

1.1. *The highest energy cosmic ray event*

The highest energy cosmic ray particle was detected by the Fly's Eye experiment [8]. We will briefly describe this event to give the reader an idea about these giant air showers. The energy of the shower is estimated to be 3×10^{20} eV. This is an enormous macroscopic energy. 10^{20} eV is equivalent to 1.6×10^8 erg, 2.4×10^{34} Hz and the energy of 170 km/h tennis ball. Fly's Eye was the first air fluorescence experiment, located in the state of Utah, U.S.A. Fig. 1 shows the shower profile of this event as measured by the Fly's Eye in the left hand panel. Note that the maximum of this shower contains more than 2×10^{11} electrons and positrons.

2. Origin of UHECR

The standard theory of cosmic ray acceleration at supernova remnants (SNR) uses astrophysical shock acceleration scenario that, in most of the recent versions, is able to reach proton energies in excess of 10^{15} eV. Heavier nuclei may be accelerated to $Z\times10^{15}$ eV. The new generation of TeV γ-ray telescopes (HESS, Magic, Kangaroo-3, Veritas) have observed TeV γ-rays associated with SNR, although there is still not a definite proof of cosmic ray acceleration.

When we discuss the acceleration of 10^{20} eV protons we have to deal with entirely different scales. These were set up by Hillas [9] from basic dimensional arguments. The first requirement for acceleration of charged nuclei in any type of object is that the magnetic field of the object contains the accelerated nucleus within the object itself. One can thus calculate a maximum theoretical acceleration energy, that does not include and efficiency factor, as

$$E_{max} \leq \gamma e Z B R \,,$$

where γ is the Lorentz factor of the shock matter, Z is the charge of the nucleus, B is the magnetic field value. and R is the linear dimension of the object.

There are very few objects that can, even before an account for efficiency, reach that energy: highly magnetized neutron stars, active galactic nuclei, gamma ray bursts, lobes of giant radiogalaxies, and possibly Gpc size shocks from structure formation.

2.1. *Possible astrophysical sources of UHECR*

In this subsection we give a brief description of some of the models for UHECR acceleration at specific astrophysical objects following Hillas' approach. For a more complete discussion one should consult a recent review paper on the astrophysical origin of UHECR [10], that contains an exhaustive list of references to particular models.

Pulsars: Young magnetized neutron stars with surface magnetic fields of 10^{13} Gauss can accelerate charged iron nuclei up to energies of 10^{20} eV [11].

The acceleration process is magnetohydrodynamic, rather than stochastic as it is at astrophysical shocks. The acceleration spectrum is very flat proportional to $1/E$. It is possible that a large fraction of the observed UHECR are accelerated in our own Galaxy. There are also models for UHECR acceleration at magnetars, neutron stars with surface magnetic fields up to 10^{15} Gauss.

Active Galactic Nuclei: As acceleration site of UHECR jets [12] of AGN have the advantage that acceleration on the jet frame could have maximum energy smaller than these of the observed UHECR by $1/\Gamma$, the Lorentz factor of the jet. The main problem with such models is most probably the adiabatic deceleration of the particles when the jet velocity starts slowing down.

Gamma Ray Bursts: GRBs are obviously the most energetic processes that we know about. The jet Lorentz factors needed to model the GRB emission are of order 100 to 1000. These models became popular with the realization that the arrival directions of the two most energetic cosmic rays coincide with the error circles of two powerful GRB. Different theories put the acceleration site at the inner [13] or the outer [14] GRB shock. To explain the observed UHECRs with GRBs one needs fairly high current GRB activity, while most of the GRB with determined redshifts are at redshift $z > 1$.

Giant Radio Galaxies: One of the first concrete model for UHECR acceleration is that of Rachen&Biermann, that dealt with acceleration at FR II galaxies [15]. Cosmic rays are accelerated at the 'red spots', the termination shocks of the jets that extend at more than 100 Kpc. The magnetic fields inside the red spots seem to be sufficient for acceleration up to 10^{20} eV, and the fact that these shocks are already inside the extragalactic space and there will be no adiabatic deceleration. Possible cosmologically nearby objects include Cen A (distance of 5 Mpc) and M87 in the Virgo cluster (distance of 18 Mpc).

Quiet Black holes: These are very massive quiet black holes, remnants of quasars, as acceleration sites [16]. Such remnants could be located as close as 50 Mpc from our Galaxy. These objects are not active at radio frequencies, but, if massive enough, could do the job. Acceleration to 10^{20} requires a mass of 10^9 $M_\odot$.

Colliding Galaxies: These systems are attractive with the numerous shocks and magnetic fields of order 20 μG that have been observed in them [17]. The sizes of the colliding galaxies are very different and with the observed high fields may exceed the gyroradius of the accelerated cosmic ray.

Clusters of Galaxies: Magnetic fields of order several μG have been observed at lengthscales of 500 Kpc. Acceleration to almost 10^{20} eV is possible, but most of the lower energy cosmic rays will be contained in the cluster forever and only the highest energy particles will be able to escape [18].

Gpc scale shocks from structure formation: A combination of Gpc scales with 1 nG magnetic field satisfies the Hillas criterion, however the acceleration at such shocks could be much too slow, and subject to large energy loss.

2.2. *Top-down scenarios*

Since it became obvious that the astrophysical acceleration up to 10^{20} eV and beyond is very difficult and unlikely, a large number of particle physics scenarios were discussed as explanations of the origin of UHECR [19]. To distinguish them from the acceleration (*bottom-up*) processes they were called *top-down*. The basic idea is that very massive (GUT scale) X particles decay and the resulting fragmentation process downgrades the energy to generate the observed UHECR. Since the observed cosmic rays have energies orders of magnitude lower than the X particle mass, there are no problems with achieving the necessary energy scale. The energy content of UHECR is not very high, and the X particles do not have to be a large fraction of the dark matter.

There are two distinct branches of such theories. One of them involves the emission of X particles by topological defects. The emission of massive X particles is possible by superconducting cosmic string loops as well as from cusp evaporation in normal cosmic strings and from intersecting cosmic strings. The X particles then decay in quarks and leptons. The quarks hadronize in baryons and mesons, that decay themselves along their decay chains. The end result is a number of nucleons, and much greater (about a factor of 30 in different hadronization models) and approximately equal number of γ-rays and neutrinos.

Another possibility is the emission of X particles from cosmic necklaces - a closed loop of cosmic string including monopoles. This particular type of topological defect has been extensively studied [20].

The other option is that the X particles themselves are remnants of the early Universe. Their lifetime should be very long, maybe longer than the age of the Universe [21]. They could also be a significant part of the cold dark matter. Being superheavy, these particle would be gravitationally attracted to the Galaxy and to the local supercluster, where their density could well exceed the average density in the Universe.

There are two main differences between bottom-up and top-down models of UHECR origin. The astrophysical acceleration generates charged nuclei, while the top-down models generate mostly neutrinos and γ-rays plus a relatively small number of protons. The energy spectrum of the cosmic rays that are generated in the decay of X particles is relatively flat, close to a power law spectrum of index α=1.5. The standard acceleration energy spectrum has index equal to or exceeding 2.

2.3. *Hybrid models*

There also models that are hybrid, they include elements of both groups. The most successful of those is the Z-burst model [22,23]. The idea is that somewhere in the Universe neutrinos of ultrahigh energy are generated. These neutrinos annihilate with cosmological neutrinos in our neighborhood and generate Z_0 bosons which decay and generate a local flux of nucleons, pions, photons and neutrinos. The resonant energy for Z_0 production is 4×10^{21} eV/m_ν(eV), where m_ν is the mass of the cosmological neutrinos. The higher the mass of the cosmological neutrinos is, the lower the resonance energy requirement. In addition, cosmological neutrinos are gravitationally attracted to concentrations of matter and their density increases in our cosmological neighborhood.

3. Propagation of UHECR

Particles of energy 10^{20} eV can interact on almost any target. The most common, and better known, target is the microwave background radiation (MBR). It fills the whole Universe and its number density of 430 cm^{-3} is large. The interactions on the radio and infrared backgrounds are also important. Let us have a look at the main processes that cause energy loss of nuclei and gamma rays.

3.1. *Energy loss processes*

The main energy loss process for protons is the photoproduction on astrophysical photon fields $p\gamma \rightarrow p + n\pi$. The minimum center of mass energy for photoproduction is $\sqrt{s_{thr}} = m_p + m_{\pi^0} \sim 1.08$ GeV. Since $s = m_p^2 + 2(1 - \cos\theta)E_p\epsilon$ (where θ is the angle between the two particles) one can estimate the proton threshold energy for photoproduction on the MBR (average energy $\epsilon = 6.3\times10^{-4}$ eV). For $\cos\theta = 0$ the proton threshold energy is $E_{thr} = 2.3 \times 10^{20}$ eV. Because there are head to head collisions and because the tail of the MBR energy spectrum continues to higher energy, the intersection cross section is non zero above proton energy of 3×10^{19} eV.

The photoproduction cross section is very well studied in accelerator experiments and is known in detail. At threshold the most important process is the Δ^+ production where the cross section reaches a peak exceeding 500 μb. It is followed by a complicated range that includes the higher mass resonances and comes down to about 100 μb. After that one observes an increase that makes the photoproduction cross section parallel to the pp inelastic cross section. The neutron photoproduction cross section is nearly identical.

Another important parameter is the proton inelasticity k_{inel}, the fraction of its energy that a proton loses in one interaction. This quantity is energy dependent. At threshold protons lose about 18% on their energy. With increase of the CM energy this fractional energy loss increases to reach asymptotically 50%.

The proton pair production [24] $p\gamma \to e^+e^-$ is the same process that all charged particles suffer in nuclear fields. The cross section is high, but the proton energy loss is of order $m_e/m_p \simeq 4 \times 10^{-4}$E. Figure 3.1 shows the energy loss length $L_{loss} = \lambda/k_{inel}$ (the ratio of the interaction length to the inelasticity coefficient) of protons in interactions in the microwave and infrared backgrounds.

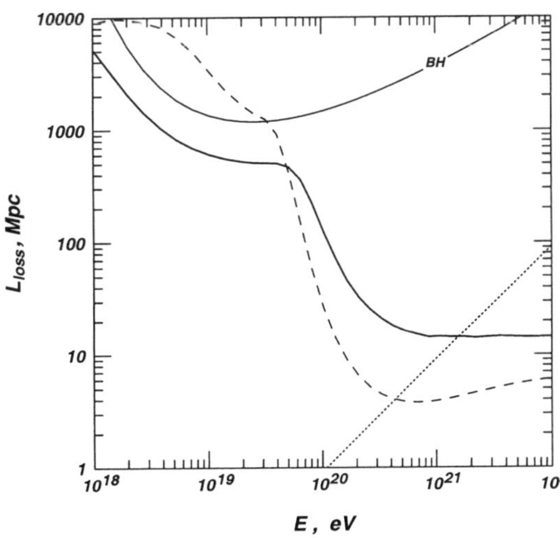

Fig. 2. Energy loss length of protons in interactions in the photon fields of the Universe.

The dashed line shows the proton interaction length and one can see the increase of k_{inel} in the ratio of the interaction to energy loss length. The contribution of the pair production is shown with a thin line. The energy loss length never exceeds 4,000 Mpc, which is the adiabatic energy loss due to the expansion of the Universe for $H_0 = 75$ km/s/Mpc. The dotted line shows the neutron decay length. Neutrons of energy less than about 3×10^{20} eV always decay and higher energy neutrons only interact.

Heavier nuclei lose energy to a different process - photodisintegration, loss of nucleons mostly at the giant dipole resonance [25]. Since the relevant energy in the nuclear frame is of order 20 MeV, the process starts at lower energy. The resulting nuclear fragment may not be stable. It then decays and speeds up the energy loss of the whole nucleus. Ultra high energy heavy nuclei, where the energy per nucleon is higher than photoproduction threshold, have also loss on photoproduction. The energy loss length for He nuclei in photodisintegration is as low as 10 Mpc at energy of 10^{20} eV. Heavier nuclei reach that distance at higher total energy.

UHE gamma rays also interact on the microwave background. The main process is $\gamma\gamma \rightarrow e^+e^-$. This is a resonant process and for interactions in the MBR the minimum interaction length is achieved at 10^{15} eV. The interaction length in MBR decreases at higher γ-ray energy and would be about a 50 Mpc at 10^{20} eV if not for the radio background. The radio background does exist but its number density is not well known.

At energies below 10^{20} eV the proton energy loss length is definitely longer than that of gamma rays. At energies above 5×10^{20} the difference is only a factor of 2, with very small energy dependence. Have in mind, though, that the flat part of the gamma ray energy loss length is due to interactions in the radio background in the 1 MHz range, which can not be detected at Earth and has to be modeled as a ratio to other astrophysical photon fields.

3.2. *Modification of the proton spectrum in propagation. Numerical derivation of the GZK effect*

Figure 3 shows in the left hand panel the evolution of the spectrum of protons because of energy loss during propagation at different distances. The thick solid lines shows the spectrum injected in intergalactic space by the source, which in this exercise is

$$\frac{dN}{dE} = A \times E^{-2}/\exp(10^{21.5}/E)\text{eV} .$$

After propagation on 10 Mpc only some of the highest energy protons are missing. This trend continues with distance and at about 40 Mpc another trend appears - the flux of protons of energy just below 10^{20} eV is above the injected one. This is the beginning of the formation of a pile-up in the range where the photoproduction cross section starts decreasing. Higher energy particles that are downgraded in this region lose energy less frequently and a pile-up is developed.

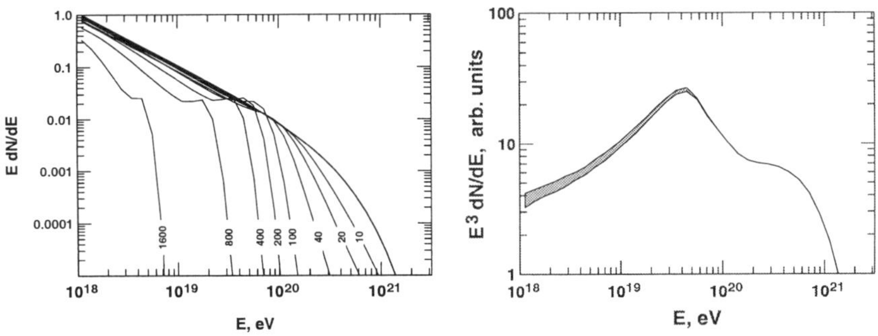

Fig. 3. Left hand panel: Evolution of the cosmic ray spectrum in propagation through different distances. Right hand panel: spectrum from homogeneous isotropic cosmic ray sources that inject spectra with $\alpha = 2$. Upper edge: cosmological evolution of UHECR sources with $n = 4$, lower one - $n = 3$.

The pile-up is better visible in the spectra of protons propagated at larger distances. One should remark that the size of the pile-up depends very strongly on the shape of the injected spectrum. If it had a spectral index of 3 instead of 2 the size of the pile-up would have been barely visible as the number of high energy particles decreased by a factor of 10.

When the propagation distance exceeds 1 Gpc there are no more particles of energy above 10^{19} eV. All these particles have lost energy in photoproduction, pair production and adiabatic losses independently of their injection energy.

In order to obtain the proton spectrum created by homogeneously and isotropically distributed cosmic ray sources filling the whole Universe one has to integrate a set of such (propagated) spectra in redshift using the cosmological evolution of the cosmic ray sources, which is usually assumed to be the same as that of the star forming regions (SFR)

$$\eta(z) \;=\; \eta(0)(1+z)^n$$

with $n = 3$, or 4 up to the epoch of maximum activity z_{max} and then either constant or declining at higher redshift. High redshifts do not contribute anything to UHECR (1600 Mpc corresponds to $z = 0.4$ for $H_0 = 75$ km/s/Mpc). After accounting for the increased source activity the size of the pile-ups has a slight increase.

The right hand panel shows the UHECR spectrum that comes from the integration of propagation spectra shown in the left hand panel with different cosmological evolutions. Obviously the importance of the cosmological evolution is very small and totally disappears for very high energy. The differential spectrum is multiplied by E^3 as is often done with experimental data to emphasize the spectral features. One can see the pile-up at 5×10^{19} eV after which the spectrum declines steeply. There is also a dip at about 10^{19} eV which is due to the energy loss on pair production which is better visible for steeper injection spectra.. These features were first pointed at by Berezinsky & Grigorieva [24].

Such should be the energy spectrum of extragalactic protons under the assumptions of injection spectrum shape, cosmic ray luminosity (4.5×10^{44} erg/Mpc^3/yr [13]), cosmological evolution and isotropic distribution of the cosmic ray sources in the Universe.

4. Production of Secondary Particles in Propagation

One interesting feature that can be used for testing of the cosmic ray injection spectrum, the cosmological evolution of the cosmic ray sources and their type and distribution in the Universe is the production of secondary particles in propagation. The energy loss of the primary protons and γ-rays is converted to secondary gamma rays, electrons, and neutrinos.

At the Δ resonance energy range 2/3 of the produced pions are neutral. Most of the energy loss (including those in e^+e^- pairs) goes to the electromagnetic component as do the muon decay electrons. The ensuing electromagnetic cascading should create a γ-ray halo around powerful UHECR sources that could be detected by the new generation of γ-ray detectors.

Since neutrinos can easily propagate from the position of their production to us they are most interesting. Cosmogenic neutrinos were first proposed by Berezinsky & Zatsepin [26] and have been since calculated many times, most recently in Ref. [27]. Every charged pion produced in a photoproduction interaction three neutrinos through its decay chain.

The spectrum of cosmogenic neutrinos extends to energies exceeding 10^{20} eV. Some currently designed and built neutrino telescopes, such as ANITA [28] are aiming at detection of cosmogenic neutrinos.

The flux and the energy spectrum of cosmogenic neutrinos also depend on the number density and the infrared background radiation (IBR) and its cosmological evolution. The interaction length of protons shown in Fig. 3.1 demonstrates that IRB does not affect much the proton spectrum after propagation because the energy loss length is longer than the one on Bethe-Heitler e^+e^- pair production. For neutrino production, however, IRB is much more important. The reason is that the IBR photons have energy much higher than MBR ones and lower energy protons can interact with them and generate pions. Optical/UV photons may have energy exceeding that of MBR by more than three orders of magnitude and even 10^{17} eV protons interact with them and generate neutrinos [29]. These neutrinos are of lower energy, proportional to the energy of the interacting protons. Even for flat E^{-2} injection spectra the number of 10^{17} eV protons exceeds that of the protons that interact in the MBR by 2.5 orders of magnitude and this compensates for the big difference in the number density of the two photon backgrounds,

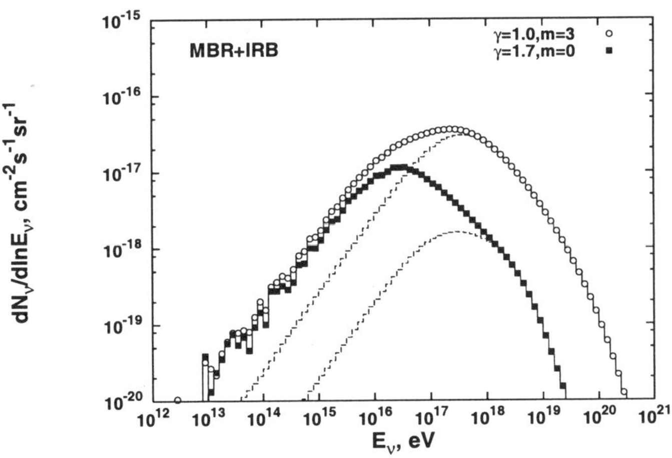

Fig. 4. Spectra of cosmogenic muon neutrinos and antineutrinos for different injection spectra and cosmological evolutions of the cosmic ray sources. Dashed histograms show the contributions from interactions in the MBR.

Fig. 4 show the spectra of cosmogenic $\nu_\mu + \bar{\nu}_\mu$ for different injection spectra and cosmogenic evolution of the cosmic ray sources. Electron neutrino spectra has a double peaked spectrum. One of the peaks, that of ν_e that are also products of π^+ decay is at the same position as the muon neutrino one. The $\bar{\nu}_e$ spectrum is generated by neutron decay and peaks at 2.5 orders of magnitude lower energy. The two peaks have about the same magnitude, about one half of the muon neutrino one in Fig. 4.

When only interactions in the MBR are considered flatter proton injection spectra generate more cosmogenic neutrinos because they contain more protons above the photoproduction threshold for the same source luminosity. Interactions in the IRB, on the other, generate more neutrinos in the case of steep spectra since the proton flux increases faster when the energy decreases.

Another important parameter is the cosmological evolution of the cosmic ray sources. Neutrinos from all redshifts reach us thus their flux reflects the highest source activity. One typical cosmological evolution of the form $(1+z)^3$ increases the flux typically by a factor of five. Cosmological models have smaller influence, but still the current favorite $\Omega_M = 0.3$ model increases the flux by almost a factor of two compared with the $\Omega_M = 1$ model. The reason is the slower expansion of the Universe at high z when Ω_Λ is accounted for.

Figure 4 compares the fluxes of cosmogenic neutrinos in two of the models of UHECR. A flat injection spectrum model [30] requires cosmological evolution at least as $(1+z)^3$ to fit UHECR spectra above a few times 10^{19} eV. The steep injection spectrum model [31] does not require any cosmological evolution to the UHECR spectrum above 10^{18} eV if the cosmic ray particles are protons with a small He component.

Before accounting for the interactions in the IRB the difference in the cosmogenic neutrino fluxes between these two models if very big as the dashed histograms in Fig. 4 show. The account for these interactions decreases the difference, although the flat injection spectrum model still generates much higher cosmogenic neutrino flux.

In case of mixed chemical composition of the high energy cosmic rays the cosmogenic neutrinos come mostly from neutron decay if the injection spectra do not extend well above 10^{22} eV [32,33]. Thus the $\bar{\nu}_e$ spectrum dominates other neutrino flavors.

Cosmogenic neutrinos are thus an important source of information about the origin of UHECR as well as for many other general astrophysical and cosmological parameters. Their fluxes are unfortunately low even in the most optimistic models and we will have to design and build special detectors in order to detect a reasonable experimental statistics.

Acknowledgments My work in the field of UHECR is funded in part by U.S. Department of energy contract DE-FG02 91ER 40626 and by

NASA grant ATP03-0000-0080. The collaboration of D. DeMarco, R. Engel, T.K. Gaisser, D. Seckel and others is highly appreciated.

References

1. J. Linsley, Phys. Rev. Lett., **10**, 146 (1963)
2. G. Cocconi, Nuovo Cimento, **3**, 1422 (1956)
3. K. Greisen, Phys. Rev. Lett. **16**, 748 (1966)
4. G.T. Zatsepin & V.A. Kuzmin, JETP Lett. **4** 78 (1966).
5. *http://www.auger.org*
6. *http://www-ta. icrr.u-tokyo.ac.jp*
7. *http://hires.physics.utah.edu*
8. D.J. Bird et al., Phys. Rev. Lett., **71**, 3401 (1993)
9. A.M. Hillas, Ann. Rev. Astron. Astrophys., **22**, 425 (1984)
10. D.F. Torres & L.A. Anchordoqui, Rep. Prog. Phys., **67**, 1663 (2004)
11. P. Blasi, R.I. Epstein & A.V. Olinto, Ap.J., **533**, L33 (2000)
12. F. Halzen & E. Zas, Ap. J., **488**, 607 (1997)
13. E. Waxman, Ap. J., **452** 1 (1995)
14. M. Vietri, Ap. J., **453**, 883 (1995)
15. J.P. Rachen & P.L. Biermann, A&A, **272**, 161 (1993)
16. E. Boldt & P. Ghosh, MNRAS, **307**, 491 (1999)
17. C.J. Cesarsky, Nucl. Phys. B (Proc. Suppl.), **28**, 51 (1992)
18. H. Kang, D. Ryu & T.W. Jones, Ap. J., **456**, 422 (1998).
19. P. Bhattacharjee & G. Sigl, Phys. Reports, **327**, 109 (2000)
20. V.S. Berezinsky& A. Vilenkin, Phys. Rev. Lett., **79**, 5202 (1997)
21. V.S. Berezinsky, M. Kahelriess & A. Vilenkin, Phys. Rev. Lett., **79**, 4302 (1997)
22. T.J. Weiler, Astropart. Phys., **11**, 303 (1999)
23. D. Fargion, B. Mele & A. Salis, Ap. J., **517**, 725 (1999)
24. V.S. Berezinsky & S.I. Grigorieva, A&A, **199**, 1 (1988)
25. J.L. Puget, F.W. Stecker & J.H. Bredekamp, Ap. J., **205**, 538 (1976)
26. V.S. Berezinsky & G.T. Zatsepin, Phys. Lett. **28b**, 423 (1969); Sov. J. Nucl. Phys. **11**, 111 (1970).
27. R. Engel, D. Seckel & T. Stanev, Phys. Rev. D64:09310 (2001)
28. *http://www.phys.hawaii.edu/ anita/web/index.htm*
29. *D. Allard et al, astro-ph/0605327*
30. J.N. Bahcall & E. Waxman, 2003, Phys.Lett.B556:1 (2003)
31. V. Berezinsky, A.Z. Gazizov & S.I. Grigorieva, Phys.Lett.B612:147 (2005)
32. M. Ave et al, Astropart. Phys. 23:19 (2005)
33. D. Hooper, A. Taylor & S. Sarkar, S., Astropart. Phys. 29:11 (2005)

GRB AS SOURCES OF ULTRA-HIGH ENERGY PARTICLES*

P. Mészáros

Dept. of Astronomy & Astrophysics and Dept. of Physics, Pennsylvania State University, University Park, PA 16802, USA
**E-mail: pmeszaros@astro.psu.edu*

In the standard gamma-ray burst model cosmic rays can be accelerated up to GZK energies $E_p \sim 10^{20}$ eV, with a flux comparable to that detected in large EAS arrays such as AUGER. Both leptonic, e.g. synchrotron and inverse Compton, as well as photomeson processes can lead to GeV-TeV gamma-rays measurable by GLAST, AGILE, or ACTs, serving as probes of the burst physics and model parameters. Photomeson interactions also produce neutrinos at energies ranging from sub-TeV to EeV, which may yield information about the fundamental interaction physics, as well as the acceleration mechanism, the nature of the sources and their environment. This emission will be probed with forthcoming experiments such as IceCube, ANITA and KM3NeT.

Keywords: Gamma-ray bursts; Ultra-high energy Cosmic Rays; High energy neutrinos; High Energy Photons

1. Introduction

The standard fireball shock model of gamma-ray bursts (GRBs) envisages the prompt MeV γ-ray production via shocks in an ultra-relativistic plasma jet moving with bulk Lorentz factors $\Gamma \gtrsim 100$) (e.g.[47]). The most obvious mechanisms responsible for the observed photons are synchrotron radiation and/or inverse Compton (IC) scattering by relativistic electrons accelerated in the shocks to a power-law distribution, although other mechanisms are also possible. The electron synchrotron spectra extend beyond 100 MeV, while IC scattering extendis into the GeV-TeV range. A significant amount of protons and neutrons may be expected in the GRB jet, along with leptons, and the protons would also be accelerated in the same shocks. This could lead to GRBs being more luminous in cosmic rays and neutrinos than in the commonly observed sub-GeV electromagnetic channels.

---

*To appear in Proceedings 2006 Erice Summer School of Cosmic Ray Astrophysics

2. Cosmic rays from GRB

The cosmic ray spectrum up to $\gtrsim 10^{15}$ eV is thought to be due to Fermi acceleration in galactic supernova remnants (SNR), and is largely made up of protons. The maximum energy for a particle of charge Z is $E_p \leq \beta ZeBR$, which is $\sim 10^{16} Z$ eV for a typical upstream magnetic field $B \sim 10^6$ G, SNR dimension $\sim 3 \times 10^{20}$ cm and $\beta_{sh} \sim 10^{-1}$. Above the "knee" near 10^{15} eV the composition is increasingly richer in heavy nuclei, and the steepening may be due to contributions from less abundant higher Z elements. This may naturally lead to an enrichment in the relative heavy element content, and a dropoff above a second knee at $\sim 10^{17}$ eV, although other explanations are also possible. The ultrahigh-energy (UHE) cosmic-rays (or UHECR) above the ankle near EeV ($= 10^{18}$ eV) energy are most probably extra-galactic. Any galactic origin at these energy, due to small magnetic deflections, would give an anisotropy of their arrival direction, contrary to the observed isotropy. The dip around $5 \times 10^{18} - 10^{19}$ eV may be due to photo-pair production of UHECR interacting with cosmic microwave (CMB) photons.[14] For UHECR above $\sim 5 \times 10^{19}$ eV, the requirement that they are not attenuated by the CMB through photo-meson ($p\gamma$) interactions constrains them to have originated within a "GZK" radius of 50-100 Mpc (e.g.[19]).

Two broad classes of UHECR models suggested are the "top-down" scenarios, which attribute UHECRs to the decay of fossil GUT defects or other primordial heavy particles, and the "bottom-up" scenarios, which assume that UHECRs are accelerated in astrophysical sources. The observed flux at Earth of UHECR of $\sim 10^{-2}$ km^{-2} year^{-1}) implies an energy injection rate into the universe of $\sim 3 \times 10^{44}$ erg Mpc^{-3} yr^{-1} above the ankle. This is similar to the $\sim$ 0.1-1 MeV γ-ray energy injection rate by the local GRBs. An problem is that, statistically, the rate of GRBs expected within a GZK radius is $\lesssim 10^{-3}$ year^{-1}. However, a plausible intergalactic magnetic field $B \sim 10^{-9}$ G with a coherence length ~ 10 Mpc will introduce a dispersion of the arrival time of $\sim 3 \times 10^7 (B/10^{-9}G)^{-2}(\lambda_B/10Mpc)$ years, which leads to the right rate of occurrence and arrival of GZK protons at Earth. Also, the same GRB shocks which are thought to accelerate the electrons responsible for the observed MeV γ-rays should also accelerate protons, and for the same conditions derived for the electrons, the maximum proton energy is $E_p \leq \beta ZeBR \gtrsim 10^{20}$ eV, i.e. GZK energies. This has motivated the conjecture that GRBs are the sources of UHECRs.[65,67] These numerical coincidences have been corroborated using new data and further considerations,[64,72,75] making GRBs promising candidates for UHECRs. Other bottom-up scenarios include active-galactic nuclei (AGNs), e.g.[14,55] and cluster shocks, e.g.[39] An unavoidable by-product of UHECR acceleration is the production of UHE neutrinos, via $p\gamma$ and pp, pn interactions.

A GRB origin is considered mostly only for the UHECR extragalactic component at $\gtrsim 10^{18.5}$ eV, with the assumption that galactic sources, such as SNRs, would be responsible for the lower energy galactic component. This is because the $\propto E^{-2}$ expected injection spectrum from GRBs (which combined with the low energy component yields approximates an effective $E^{-2.7}$ in the sub-GZK range) has neither the spectrum nor the energy density, under normal propagation conditions, to explain the PeV to EeV flux observed. However, a GRB origin model has been proposed,[75] based on modified cosmic ray transport and confinement times in the galaxy, which attributes the entire PeV to GZK energy range to GRB cosmic rays.

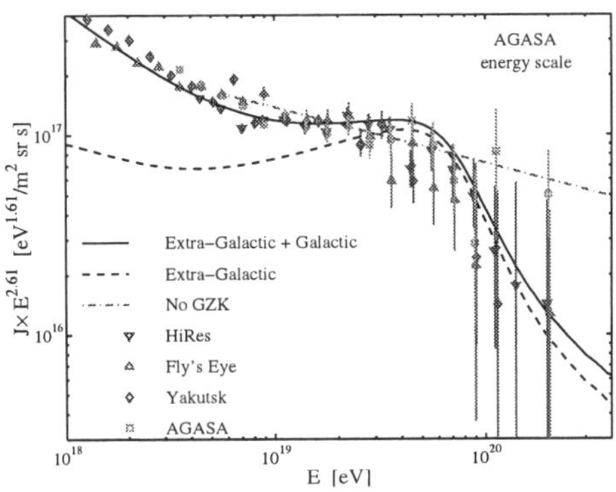

Fig. 1. Comparison of UHECR data with the predictions of a model where extra-galactic protons in the energy range $E_p \leq 10^{21}$ eV are produced by cosmologically distributed sources at a rate and spectrum expected from GRB, $E_p^2 \frac{d\dot{N}_p}{dE_p} = 0.65 \times 10^{44}$erg Mpc^{-3} yr$^{-1}\phi(z)$.[73]

The most commonly discussed version of the GRB bottom-up scenario considers the UHECR to be protons accelerated in GRB internal shocks,[67,68,72] while another version attributes them to acceleration in external shocks.[25,64,65] A caveat is that the internal shock scenario relies on the assumption that GRB prompt gamma-ray emission is due to internal shocks. Although this is the leading scenario, there is no strong proof so far for this, as there is for external shocks (e.g., there are efficiency and spectrum issues, etc.). On the other hand, a Poynting flux dominated GRB model would have to rely on magnetic dissipation and reconnection, accelerating electrons and hence also accelerating protons- but details remain to be investigated. The external shock model would have to rely on a magnetized medium[64] to reach the desired cosmic ray energy.

A direct confirmation of a GRB (or other) origin of UHECRs will be difficult. The next generation cosmic ray detectors such as the *Pierre Auger Observatory*[80] will have a substantially enhanced effective target area, which will greatly improve the cosmic ray count statistics. This will help to disentangle the two scenarios (top-down or bottom-up) and will reveal whether a GZK feature indeed exists. Within the bottom-up scenario, the directional information may either prove or significantly constrain the alternative AGN scenario, and may eventually shed light on whether GRBs are indeed the sources of UHECRs.

3. GeV and TeV γ-ray emission from GRB

The evidence for cosmic ray acceleration up to $\sim 10^{15}$ eV in SNRs is becoming stronger, based on TeV γ-ray observations by imaging air Cherenkov telescopes (IACTs) such as HESS. The observed TeV γ-ray spectrum is independent of location and azimuth in well-resolved SNR images, as expected from an origin in pion decay created by protons with long mean-free path. An electromagnetic origin such as IC scattering by electrons would result in a varying spectrum depending on local magnetic field strength, because of a shorter mean-free path, e.g.[1,30]

Long GRBs are increasingly being found associated with supernovae (e.g.[44,61]). If GRBs also accelerate cosmic rays, then these CRs could leave long-lasting UHE photon signatures in supernova remnants associated with GRBs. One example may be the SN remnant W49B, which may be a GRB remnant. A signature of a neutron admixture in the relativistic cosmic ray outflow would be a TeV γ-ray signature due to IC scattering following neutron decay, on timescales of thousands of years after the burst.[40] Another possible GRB supernova remnant considered is the unidentified TeV source HESS J1303-631.[9] The imaging of the surrounding emission at GeV and TeV energies could provide new constraints on the GRB jet structure.

While the GRB remnants could be essentially steady γ-ray sources, the prompt GRB emission is also thought to result in GeV-TeV photons. The electrons accelerated by the internal and/or external shocks via the Fermi mechanism, in a turbulently enhanced magnetic field, have a power-law energy distribution, leading to electron synchrotron radiation which in the observer frame extends beyond 100 MeV. Inverse Compton (IC) scattering of such synchrotron photons leads to GeV to TeV spectral com-

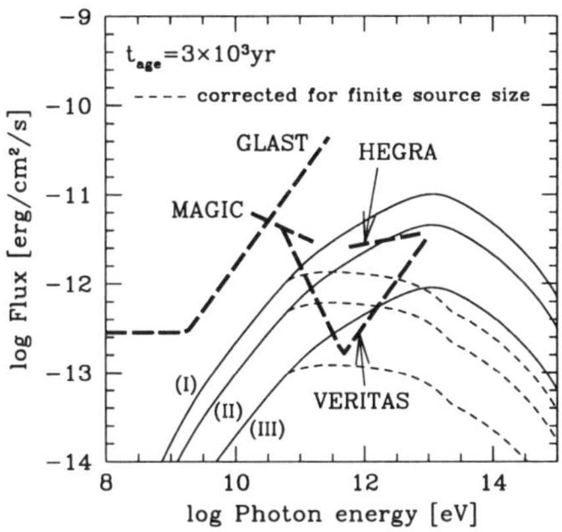

Fig. 2. The photon flux from β-decay electrons in W49.[40] Solid lines are the IC scattering of CMB, IR and optical photons. Bold long dashed lines are the sensitivities of GLAST, HEGRA and VERITAS. Dashed lines are the flux of β-decay emission multiplied by $(\Omega/\pi\theta_{\rm cut})^{-1/2}$, where Ω is solid angle of the emitting region and $\theta_{\rm cut} \sim 0.1°$ is angular cut.

ponents.[17,24,32,45,51,53] While the emission can extend to TeV energies, such photons are likely to be degraded to lower energies by $\gamma\gamma$ pair production, either in the source itself[45,51,57] or (unless the GRB is at very low redshifts) in the intervening intergalactic medium.[18,21] The internal $\gamma\gamma$ pair production leads to interesting limits on the bulk Lorentz Γ factor of the GRB outflow. This is because the cutoff in the observer frame is energy dependent: the Lorentz transformation to the CM system changes the threshold to $2m_ec^2(1-\cos\theta)$, where θ is the relative angle of incidence of the two photons, and causality imposes the condition $\theta \lesssim 1/\Gamma$ (also, the optical depth $\tau_{\gamma\gamma} \sim \sigma_T(r/\Gamma)(L/4\pi r^2\Gamma^2 m_ec^3)$ depends on Γ). This cutoff sensitivity to Γ provides a diagnostic for the bulk Lorent factor.[11,12,43]

Another photon emission mechanism at these energies could be π^0 decay following $p\gamma$ photomeson interactions between shock-accelerated protons and MeV or other photons in the GRB shock region.[16,31,63] This can be important, provided a) there are protons in the outflow and they are accelerated in significant numbers; b) the relativistic proton energy exceeds by at least one order of magnitude the energy in relativistic electrons or in γ-rays, and c) the proton spectral index is hard, e.g. -2, rather than -2.2 or softer; otherwise, both the proton synchrotron and the $p\gamma$ components can be shown to be less important at GeV-TeV energies than the

IC component.[76] The p, γ interactions in the GRB outflow will lead to both hadronic (producing charged and neutral pions) and electromagnetic cascades (muons and electron pairs) and neutrinos. Photons will be produced by proton, muon and secondary electron synchrotron and IC losses. One of the characteristics of such hadronic mechanisms involving electromagnetic cascades is that since the proton losses are slower than those of electrons, the afterglow predicted by proton cascades stretches over a longer timescale.[16] The GeV light curves arising from such hadronic mechanisms would have a different shape from those from leptonic mechanisms such as primary electron synchro-Compton.[26,27] One also expects cascades to have a harder spectrum than primary synchro-Compton, which is the basis of the argument for hadronic cascades in GRB 941017 made by.[28,34] It can however be argued that under some conditions a purely leptonic (primary electron) synchro-Compton mechanisms can explain the same observations too,[35,52] so the hadronic identification is inconclusive.

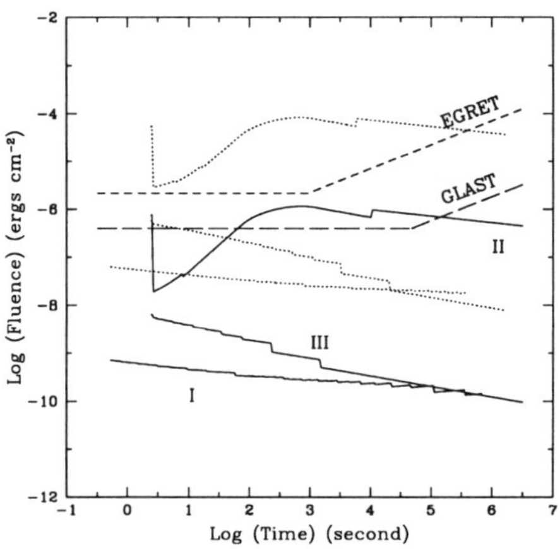

Fig. 3. The 400 MeV-200 GeV band νF_ν lightcurves,[76] for GRB dominated by proton synchrotron (I), electron IC (II) and electron synchrotron (III) located at cosmological distances $z = 1$ (solid) and $z = 0.1$ (dotted). The electron IC component gives an extended duration GeV emission easily detectable by *GLAST* at $z = 1$ in a regime II burst (solid line).

A hadronic GeV photon component can also be expected in a baryonic GRB outflow since neutrons are likely to be present, and when these decouple from the protons, before any shocks occur, pn inelastic collisions will lead to pions, including π^0, resulting in UHE photons which cascade down to the GeV range.[10,23,60] The final GeV spectrum results from a complex cascade, but a rough estimate indicates that a 1-10 GeV flux should be detectable with GLAST for $z \lesssim 0.1$.[10]

The recent detection of delayed X-ray flares during the afterglow phase of gamma-ray bursts (GRBs) with the Swift satellite (e.g.[48,50,77]) suggests an inner-engine origin of these flares, at radii inside the deceleration radius characterizing the beginning of the forward shock afterglow emission. Given the observed temporal overlapping between the flares and afterglows, one expects an inverse Compton (IC) emission arising from such flare photons scattered by forward shock afterglow electrons.[66] This IC emission would produce GeV-TeV flares, which may be detected by GLAST and ground-based TeV telescopes. The detection of GeV-TeV flares combined with low energy observations may help to constrain the poorly known magnetic field in afterglow shocks.

Photons up to $\lesssim$18 GeV energies have been observed[38] in at least four GRBs[29] with the EGRET experiment on CGRO. At higher energies, a tentative $\gtrsim$0.1 TeV detection at the 3σ level of GRB970417a has been reported with the water Cherenkov detector Milagrito.[7] An analysis of recent TeV upper limits with the Milagro array is given by.[8] Another possible TeV detection[54] of GRB971110 has been reported with the GRAND array, at the 2.7σ level. Stacking of data from the TIBET array for a large number of GRB time windows has led to an estimate of a $\sim 7\sigma$ composite detection significance.[5] Better sensitivity is expected from upgrades to MILAGRO, as well as from atmospheric Cherenkov telescopes opertaing or under construction such as HESS, MAGIC, CANGAROO-III and VERITAS.[49,74] The MAGIC telescope has the ability to slew in less than 30 seconds to any location, and has been responding to GCN alerts from Swift in search of prompt TeV emission, yielding upper limits (e.g.[3] However, GRB detections in the TeV range are expected only for rare nearby events, since at this energy the mean free path against $\gamma\gamma$ absorption on the diffuse IR photon background is $\sim$ few hundred Mpc.[18,21] The mean free path is much larger at GeV energies, and several dozens should be detectable with satellites such as AGILE,[78] and hundred with large area satellites such as GLAST.[76,81]

4. High energy neutrinos

High energy neutrinos in the $\sim 10^{11} - 10^{17}$ eV range, detectable by experiments such as IceCube or KM3NeT, may be produced in GRBs in a way similar to the beam-dump experiments in particle accelerators. Shock-accelerated protons interacting with ambient radiation and/or plasma material by photonuclear ($p\gamma$) and/or inelastic nuclear (pp/pn) collisions produce charged pions ($\pi^{\pm}$) and neutral pions (π^0). Neutrinos are produced from $\pi^{\pm}$ decays along with muons and electrons. Such neutrinos may serve as diagnostics of the presence of relativistic shocks, and as probes of the acceleration mechanism and the magnetic field strength. The flux and spectrum of EeV neutrinos depends on the density of the surrounding gas, while the TeV-PeV neutrinos depend on the fireball Lorentz factor. Hence, the detection of very high energy neutrinos would provide crucial constraints on the fireball parameters and GRB environment. Lower energy ($\lesssim$ TeV) neutrinos originating from sub-stellar shocks, on the other hand, may provide useful information on the GRB progenitor.

4.1. *Neutrinos contemporaneous with the gamma-rays*

Shock accelerated protons interact dominantly with observed synchrotron photons with ~MeV peak energy in the fireball to produce a Δ^+ resonance as $p\gamma \to \Delta^+$. The threshold condition to produce a Δ^+ is $E_p E_\gamma = 0.2\Gamma_i^2$ GeV^2 in the observer frame, which corresponds to a proton energy of $E_p = 1.8 \times 10^7 E_{\gamma,\mathrm{MeV}}^{-1} \Gamma_{300}^2$ GeV. The short-lived Δ^+ decays either to $p\pi^0$ or to $n\pi^+ \to n\mu^+\nu_\mu \to ne^+\nu_e\bar{\nu}_\mu\nu_\mu$ with roughly equal probability. It is the latter process that produces high energy neutrinos in the GRB fireball, contemporaneous with the γ-rays.[69] The secondary π^+ receive $\sim 20\%$ of the proton energy in such an $p\gamma$ interaction and each secondary lepton roughly shares 1/4 of the pion energy. Hence each flavor (ν_e, $\bar{\nu}_\mu$ and ν_μ) of neutrino is emitted with $\sim 5\%$ of the proton energy, dominantly in the PeV ($= 10^{15}$ eV) range, with equal ratios. The diffuse muon neutrino flux from GRB internal shocks due to proton acceleration and subsequent $p\gamma$ interactions is shown as the short dashed line in Fig. 4. The flux is compared to the Waxman-Bahcall limit of cosmic neutrinos from optically thin sources, which is derived from the observed cosmic ray flux.[71] The fluxes of all three neutrino flavors (ν_e, ν_μ and ν_τ) are expected to be equal after oscillation in vacuum over astrophysical distances.

4.2. *Neutrinos from GRB afterglows*

The GRB afterglow arises when relativistic plasma jet or outflow starts being slowed down by the external medium (e.g. the interstellar medium, ISM), driving a blast wave ahead of the jet. This produces an external forward shock or blast wave, and a reverse shock in the jet. The external

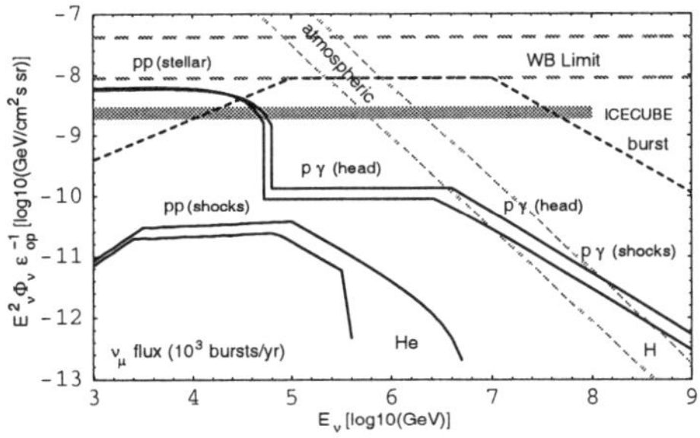

Fig. 4. Diffuse ν_μ flux arriving simultaneously with the γ-rays in observed GRB (dark short-dashed curve), compared to the Waxman-Bahcall (WB) diffuse cosmic ray bound (light long-dashes) and the atmospheric neutrino flux (light short-dashes). Also shown is the diffuse muon neutrino precursor flux (solid lines) from sub-stellar jet shocks in two GRB progenitor models, with stellar radii $r_{12.5}$ (H) and r_{11} (He). These neutrinos would be present also in electromagnetically dark bursts (choked jets), and would arrive 10-100 s before the γ-rays in electromagnetically detected bursts.

shock takes place at a radius $r_e \sim 4\Gamma_e^2 c\Delta t \sim 2\times 10^{17}\Gamma_{250}^2\Delta t_{30}$ cm which is well beyond the internal shock radius.[71] Here $\Gamma_e \approx 250\Gamma_{250}$ is the bulk Lorentz factor of the ejecta after the partial energy loss from emitting γ-rays in the internal shocks, and $\Delta t = 30\Delta t_{30}$ s is the duration of the GRB jet. Neutrinos are produced in the external reverse shock due to $p\gamma$ interactions of shock accelerated protons predominantly with synchrotron soft x-ray photons produced by electrons. The energy of the neutrinos from the afterglow would be in the EeV range as more energetic protons interact with these soft photons to produce Δ^+. The efficiency of proton to pion conversion by $p\gamma$ interactions in the external shocks (afterglow) is typically smaller than in the internal shocks because $r_e \gg r_i$, implying lower photon density.

In the case of a massive star progenitor the GRB jet may be expanding into a stellar wind much denser than the typical ISM density of $n \simeq 1$ cm^{-3}, which is emitted by the progenitor prior to its collapse. For a wind with mass loss rate of $\sim 10^{-5} M_\odot$ yr^{-1} and velocity of $v_w \sim 10^3$ km/s, the wind density at the typical external shock radius would be $\simeq 10^4$ cm^{-3}. The higher density implies a lower Γ_e, and hence a larger fraction of proton energy lost to pion production. Protons of energy $E_p \gtrsim 10^{18}$ eV lose all their energy to pion production in this scenario producing EeV neutrinos.[20]

4.3. *Precursor neutrinos*

In long duration GRBs, the relativistic jet is launched near the central black hole resulting from the collapse of the stellar core, deep inside the star. As the jet burrows through the star, it may or may not break through the stellar envelope.[46] Internal shocks in the jet, while it is burrowing through the star, can produce high energy neutrinos due to accelerated protons, dominantly below ~ 10 TeV, through pp and $p\gamma$ interactions.[58] The jets which successfully penetrate through the stellar envelope result in GRBs (γ-ray bright bursts), while the jets which choke inside the stars do not produce GRBs (γ-ray dark bursts). However, in both cases high energy neutrinos can be produced in the internal shocks, which slice through the stellar envelope since they interact very weakly with matter.

These neutrinos from the relativistic buried jets are emitted as precursors ($\sim$ 10-100 s prior) to the neutrinos emitted from the GRB firebal, in the case of an electromagnetically observed burst. In the case of a choked (electromagnetically undetectable) burst, no direct detection of neutrinos from individual sources is possible. However the diffuse neutrino signal is boosted up in both scenarios. The diffuse neutrino flux from two progenitor star models are shown in Fig. 4, one for a blue super-giant (labeled H) and the other a Wolf-Rayet type (labeled He). The neutrino component which is contemporaneous with the γ-ray emission (i.e. which arrives after the precursor) is shown as the dark dashed curve, and is plotted assuming that protons lose all their energy to pions in $p\gamma$ interactions in internal shocks. (For precursor neutrinos in supranova models see[33]).

4.4. *GRB-Supernova connection*

A fraction of long GRBs have recently been shown to be associated with supernovae of type Ib/c.[22] A GRB jet loaded with baryons would then leave long-lasting UHECR, neutrino and photon signatures in those supernova remnants which were associated with a GRB at the time of their explosion. Examples of possible VHE photon signatures discussed in §3 include the SN remnant W49B[40] and HESS unidentified sources.[9] The GRB related UHECR in such sources would lead also to UHE neutrons, whose delayed decay would give rise to TeV neutrinos. Cosmic-rays accelerated in the SNR-GRB remnant, which may be similar to SNRs observed as TeV γ-ray sources such as RX J1713.7-3946,[4] would also be expected to produce UHE neutrinos. The energy of the neutrinos and γ-rays originating from $p\gamma$ and/or pp/pn interactions would be higher in the case of GRB remnants compared to common SNRs because of the higher expansion velocity.

4.5. *Neutrino flavor astrophysics*

High energy neutrinos from astrophysical optically thin sources are expected to be produced dominantly via $p\gamma$ interactions. The subsequent decay of π^+, and neutrino flavor oscillations in vacuum, lead to an observed anti-electron to total neutrino flux ratio of $\Phi_{\bar{\nu}_e} : \Phi_\nu \simeq 1 : 15$.[42] At high energies this ratio may be lower even,[41] since the muons suffer significant electromagnetic energy loss prior to decay.[56] In the case of pp/pn interactions, typically attributed to optically thick sources, $\pi^\pm$ are produced in pairs and the corresponding expected flux ratio on Earth is $\Phi_{\bar{\nu}_e} : \Phi_\nu \simeq 1 : 6$. However even in optically thin sources the nominal $\Phi_{\bar{\nu}_e} : \Phi_\nu$ ratio may be enhanced above $1 : 15$ by $\gamma\gamma \to \mu^\pm$ interactions and subsequent $\mu^\pm$ decays.[59] The targets are the usual synchrotron photons, while UHE incident photons are provided by the $p\gamma \to p\pi^0 \to p\gamma\gamma$ channel itself. This mechanism yields an enhancement ratio $\Phi_{\bar{\nu}_e} : \Phi_\nu \simeq 1 : 5$ solely from $\mu^\pm$ decays. Measurement of the $\bar{\nu}_e$ to ν flux ratios may be possible by IceCube through the Glashow resonant interaction $\bar{\nu}_e e \to W^- \to$ anything at $E_\nu \simeq 6.4$ PeV.[6] Any enhancement over the $1 : 15$ ratio, e.g., from a single nearby GRB would then suggest a $\gamma\gamma$ origin. In fact, the flux of $\gamma\gamma$ neutrinos depends on source model parameters such as magnetization, radius etc. A calculation of the $\bar{\nu}_e$ to ν flux ratio,[59] including the $p\gamma$ and $\gamma\gamma$ channels from a GRB internal shocks with different model parameters, shows that the ratio is enhanced from the $p\gamma$ value of $1/15$ in the small energy range where $\gamma\gamma$ interactions contribute significantly. This may be used to diagnose the GRB model parameters.

5. Conclusions

The leading GRB photon radiation scenario, the fireball shock model, is well supported as far as the properties of the external shock, and is expected to be a strong source of GeV γ-rays. The TeV component may be observable in nearby bursts, providing important contraints on the burst physics and the intergalactic IR background. Other aspects of GRB models, such as internal and reverse shocks, are so far more ambiguous. The detection of high energy neutrino emission from GRBs would serve as a direct test for this, as well as for a baryonic jet component, where the bulk of the energy is carried by baryons. On the other hand, an alternative Poynting flux dominated GRB jet model would have to rely on magnetic dissipation and reconnection, accelerating electrons and hence also accelerating protons– but there would be much fewer protons to accelerate and probably to much lower energy.

The *Pierre Auger Cosmic Ray Observatory* currently under construction will have a very large area ($\sim$ 3000 km^2 each in its Southern and Northern hemisphere locations),[80] greatly improving the UHECR count statistics. This will help disentangle the competing top-down and bottom-up scenarios, and will reveal whether a GZK feature indeed exists. Within the bottom-up scenario, the directional information may either prove or significantly constrain the alternative AGN scenario, and may eventually shed light on whether GRBs are indeed sources of UHECRs.

Upcoming experiments such as IceCube,[83] ANITA,[79] KM3NeT[82] and Auger[80] are currently being built to detect high energy astrophysical neutrinos. They can provide very useful information on the particle acceleration, radiation mechanism and magnetic fields, as well as about the sources and their progenitors. Direct confirmation of a GRB origin of UHECRs is difficult but the highest energy neutrinos may indirectly serve that purpose pointing directly back to their sources. Most GRBs are located at cosmological distances (with redshift $z \sim 1$) and individual detection of them by km scale neutrino telescopes may not be possible. The diffuse neutrino flux is then dominated by a few nearby bursts. The likeliest prospect for UHE ν detection is from these nearby GRBs in correlation with electromagnetic detection.

The prospect for high energy neutrino astrophysics is very exciting, with AMANDA[13,62] and RICE[15] already providing useful limits on the diffuse flux from GRBs and with IceCube[2,37] on its way. The detection of TeV and higher energy neutrinos from GRBs would be of great importance for understanding the astrophysics of these sources such as the hadronic vs. the magnetohydrodynamic composition of the jets, as well as the CR acceleration mechanisms involved. High energy neutrinos from GRBs may also serve as probes of the highest redshift generation of star formation in the Universe, since they can travel un-attenuated, compared to the conventional electromagnetic astronomical probes.

I am grateful to S. Razzaque and X-Y. Wang for collaborations. Work supported by NSF AST0307376 and NASA NAG5-13286.

References

1. Aharonian, F., et al., 2006, Astron. Ap
2. Ahrens, J., et al., 2004, New Astron. Rev., 48:519
3. Albert, J., et al., 2006, ApJ, 641:L9

4. Alvarez-Muñiz, J. and Halzen, F., 2002, ApJ, 576:L33
5. Amenomori, M., et al., 2001, AIPC, 558:844
6. Anchordoqui, L. A., Goldberg, H., Halzen, F. and Weiler, T. J., 2005, Phys. Lett. B621:18
7. Atkins, R., et al., 2000, ApJ, 533:L119
8. Atkins, R., et al., 2005, ApJ, 630:996
9. Atoyan, A., Buckley, J. and Krawczynski, H., 2006, ApJ 642:L43
10. Bahcall, J. N. and Mészáros, P., 2000, Phys. Rev. Lett., 85:1362
11. Baring, M. and Harding, A., 1997, ApJ, 491:663
12. Baring, M., 2006, ApJ, in press (astro-ph/0606425)
13. Becker, J., et al., 2006, Astropart. Phys., 25:118
14. Berezinsky, V. S., Gazizov, A. Z. and Grigorieva, S. I., 2005, Phys. Lett. B, 612:147
15. Besson, D. Z., Razzaque, S., Adams, J. and Harris, P., Astropart. Phys., submitted, (astro-ph/0605480)
16. Böttcher, M. and Dermer, C., 1998, ApJ, 499:L131
17. Chiang, J. and Dermer, C., 1999, ApJ, 512:699
18. Coppi, P. and Aharonian, F., 1997, ApJ, 487:L9
19. Cronin, J., 2005, Nucl. Phys. Proc. Suppl., 138:465-491 (astro-ph/0402487)
20. Dai, Z. G. and Lu, T., 2001, ApJ, 551:249
21. de Jager, O. C. and Stecker, F. W., 2002, ApJ 566:738
22. Della Valle, M, A.A., in press (astro-ph/0504517)
23. Derishev, E. V., Kocharovsky, V. V. and Kocharovsky, Vl. V., 1999, ApJ 521:640
24. Dermer, C., Chiang, J. and Mitman, K., 2000, ApJ, 537:785
25. Dermer, C., 2005, in Proc. "Gamma Ray Bursts in the Swift Era", Washington, D.C., eds. S. Holt, et al, AIPC, in press
26. Dermer, C. and Atoyan, A., 2003, Phys. Rev. Let., 91:1102
27. Dermer, C. and Atoyan, A., 2004, Astron. Ap. 418:L5
28. Dermer, C. and Atoyan, A., 2004b, AIPC 727:557
29. Dingus, B., 2003, AIPC 662:240
30. Enomoto, R., et al., 2002, Nature 416:823
31. Fragile, P., et al., 2004, Astropart. Phys., 20:598
32. Guetta, D & Granot, J, 2003, ApJ 585:885
33. Guetta, D & Granot, J, 2003c, PRL 90:191102
34. Gonzlez, M. M., et al., 2003, Nature, 424:749
35. Granot, J. and Guetta, D., 2003, ApJ, 598:L11
36. Hoerandel, J. R., et al., 2005, AIPC 801:72
37. Hulth, P. O., in *NO-VE 2006, Neutrino Oscillations in Venice*, Italy (astro-ph/0604374).
38. Hurley, K., et al., 1994, Nature, 372:652
39. Inoue, S., Aharonian, F. and Sugiyama, N., 2005, ApJ 628:L9
40. Ioka, K., Kobayashi, S. and Mészáros, P., 2004, ApJ 613:L171
41. Kashti, T. and Waxman, E., 2005, Phys. Rev. Lett., 95:181101
42. Learned, J. G. and Pakvasa, S., 1995, Astropart. Phys., 3:267
43. Lithwick, Y. and Sari, R., 2001, ApJ, 555:540

44. Masetti, N., Palazzi, E., Pian, E. and Patat, F., 2006, 6GCN 4803
45. Mészáros, P., Rees, M. J. and Papathanassiou, H., 1994, ApJ 432:181
46. Mészáros, P. and Waxman, E., 2001, Phys. Rev. Lett., 87:171102
47. Mészáros, P., 2006, Rep. Prog. Phys. 69:2259-2321 (astro-ph/0605208)
48. Nousek, J., et al., 2006, ApJ, in press (astro-ph/0508332)
49. Ong, R., 2005, in Procs. ICRC 2005, in press (astro-ph/0605191)
50. Panaitescu, A., et al., 2006a, MNRAS, in press (astro-ph/0508340)
51. Papathanassiou, H. and Mészáros, P., ApJ, 471:L91
52. Pe'er, A. and Waxman, E., 2004, ApJ 603:L1
53. Pe'er, A. and Waxman, E., 2004b, ApJ 613:448
54. Poirier, J., et al., 2003, Phys. Rev. D, 67:2001
55. Rachen, J. and Biermann, P., 1993, Astron. Ap., 272:161
56. Rachen, J. P. and Mészáros, P, 1998, Phys. Rev. D, 58:123005
57. Razzaque, S., Mészáros, P. and Zhang, B., 2004, ApJ 613:1072
58. Razzaque, S., Mészáros, P. and Waxman, E., 2003, Phys. Rev. D, 68:083001
59. Razzaque, S., Mészáros, P. and Waxman, E., 2006, Phys. Rev. D, 73:103005
60. Rossi, E., Beloborodov, A. and Rees, M. J., 2005, MNRAS, submitted (astro-ph/0512495)
61. Stanek, K., et al., 2003, ApJ, 591:L17
62. Stamatikos, M., et al., 2004, AIPC 727:146
63. Totani, T., 1999, ApJ, 511:41
64. Vietri, M., de Marco, D. and Guetta, D., 2003, ApJ 592:378
65. Vietri, M., 1995, ApJ 453:883
66. Wang, X. Y., Li, Z. and Mészáros, P., 2006, ApJ, 641:L89
67. Waxman, E., 1995, Phys. Rev. Lett., 75:386
68. Waxman, E., 1995b, ApJ, 452:L1
69. Waxman, E. and Bahcall, J. N., 1997, Phys. Rev. Lett., 78:2292
70. Waxman, E. and Bahcall, J. N., 1999, Phys. Rev. D, 59:023002
71. Waxman, E. and Bahcall, J. N., 2000, ApJ, 541:707
72. Waxman, E., 2004, ApJ, 606:988
73. Waxman, E., 2005, Phys. Scripta, T121:147-152
74. Weekes T., 2006, in Procs. Energy Budget of the High Energy Universe; Kashiwa, Japan, Feb 22-24, 2006 (astro-ph/0606130)
75. Wick, S., Dermer, C. and Atoyan, A., 2004, Astropart. Phys., 21:125
76. Zhang, B. and Mészáros, P., 2001b, ApJ, 559:110
77. Zhang, B., et al., 2006, ApJ, in press (astro-ph/0508321)
78. AGILE: http://agile.rm.iasf.cnr.it/doc/a-science-27.pdf
79. ANITA: http://www.ps.uci.edu/ anita/
80. AUGER : http://www.auger.org/
81. GLAST: http://glast.gsfc.nasa.gov/
82. KM3NeT: http://km3net.org/
83. ICECUBE: http://icecube.wisc.edu/

ORIGIN AND PHYSICS OF THE HIGHEST ENERGY COSMIC RAYS: WHAT CAN WE LEARN FROM RADIO ASTRONOMY?

Peter L. Biermann*[1,2,3], P. Gina Isar [4], Ioana C. Maris[4], Faustin Munyaneza[1], & Oana Taşcău[5]

[1] *Max-Planck Institute for Radioastronomy, Bonn, Germany*
[2] *Department of Physics and Astronomy, University of Bonn, Bonn, Germany,*
[3] *Department of Physics and Astronomy, University of Alabama, Tuscaloosa, AL, USA*
[4] *FZ Karlsruhe, and Physics Dept., Univ. of Karlsruhe*
[5] *University of Wuppertal, Wuppertal, Germany*

Here in this lecture we will touch on two aspects, one the new radio methods to observe the effects of high energy particles, and second the role that radio galaxies play in helping us understand high energy cosmic rays. We will focus here on the second topic, and just review the latest developments in the first. Radio measurements of the geosynchrotron radiation produced by high energy cosmic ray particles entering the atmosphere of the Earth as well as radio Čerenkov radiation coming from interactions in the Moon are another path; radio observations of interactions in ice at the horizon in Antarctica is a related attempt. Radio galaxy hot spots are prime candidates to produce the highest energy cosmic rays, and the corresponding shock waves in relativistic jets emanating from nearly all black holes observed. We will review the arguments and the way to verify the ensuing predictions. This involves the definition of reliable samples of active sources, such as black holes, and galaxies active in star formation. The AUGER array will probably decide within the next few years, where the highest energy cosmic rays come from, and so frame the next quests, on very high energy neutrinos and perhaps other particles.

Keywords: Cosmic rays, magnetic fields, active galactic nuclei, black holes, radio galaxies

1. Introduction

Recent years have seen a proliferation in new experimental efforts to measure very high energy particles, both in actually constructing huge new arrays like AUGER in Argentina, but also new attempts to measure high energy particles in new ways, mostly focussing on the radio range. We can-

*E-mail:plbiermann@mpifr-bonn.mpg.de

not but mention in passing the hugely successful telescopes for TeV gamma rays, like HESS, MAGIC, MILAGRO and Cangaroo, as well as the neutrino observatory IceCube for high energy neutrinos. Here we focus on yet higher energies, near and beyond EeV (= 10^{18} eV). For the subject covered in this lecture, the reader is refered to important textbooks.[1–3]

On the other hand, our theoretical understanding of possible production of high energy charged particles, neutrinos and photons is also reaching a measure of maturity, with many efforts concentrating on the role of shock waves and flows in relativistic jets, such as in radio galaxy hot spots as discussed in various review articles.[4–8]

2. Radio detection methods

It has been recognized many decades ago, that high energy particles produce secondary radio emission. The emission is now far along to be understood as geosynchrotron emission,[9–11] or as radio Čerenkov emission. Now we have established efforts under way to observe, calibrate and use such emissions to set limits, possibly soon measure, high energy neutrinos, very high energy cosmic rays and also unknown particles.

The furthest along has been the effort to use geosynchrotron emission, when the airshower is a directly visible radio spot in the sky. The observation of high energy cosmic rays will be incorporated into the LOFAR array, in the Netherlands; the LOPES array in Karlsruhe, Germany, undertakes the control, and calibration of these emissions.[12–21] Radio Čerenkov emission from the Moon is another effort, as is the corresponding observation in ice at the horizon in Antarctica.[22–24] There are corresponding efforts elsewhere, and also some tests have been done to use salt-domes.[25]

3. Active galactic nuclei

For Galactic cosmic rays clearly supernova explosions are the prime candidates to be sources of cosmic rays; the latest results from the TeV telescopes strongly support this expectation and interpretation. The energy range to which cosmic rays derive from Galactic sources is not entirely certain, but the transition to extragalactic cosmic rays is somewhere near $3\,10^{18}$ eV, perhaps at slightly lower energy.

Where ultra high energy cosmic rays come from is not certain, but the most promising candidates are radio galaxy hot spots and other shocks in relativistic jets, and gamma ray bursts, almost certainly also a phenomenon involving ultra-relativistic jets.

Here we wish to concentrate on radio galaxies, as in their case the argument is persuasive, that they accelerate cosmic rays to extremely high energy. This does not necessarily imply that they are the sources for the events we observe.

4. Radio galaxies

Radio galaxy hot spots as well as many other knots in radio jets show an ubiquitous cutoff near $3\,10^{14}$ Hz, originally discovered in the mid 1970ies. Radio, infrared and optical observations strongly suggest that these hot spots and knots are weakly relativistic shock waves. The basic phenomenon seems to be quite independent of external circumstances, and so requires a very simple mechanism. It has been shown that in such shocks protons can be accelerated,[26] subject to synchrotron and Inverse Compton losses, then initiate a wave-field in the turbulent plasma, and the electrons scattered and accelerated in the shock are then limited to emit at that synchrotron frequency observed, $\nu^{\star} \simeq 3\,10^{14}$ Hz, almost independent of details.

4.1. *The maximum energy*

The maximum energy of the protons $E_{p,max}$ from this loss limit implicated can then be written as

$$E_{p,max} \simeq 1.5\,10^{20}\,\mathrm{eV}\,\left(\frac{\nu_e^{\star}}{3.10^{14}\,\mathrm{Hz}}\right)^{1/2}\,B^{-1/2} \tag{1}$$

Here we are independent of all the detailed assumptions about the intensity of the turbulence, and the exact shock speed; the dependence through the magnetic field on other parameters is only with the 1/7-power; typical magnetic field strength inferred are between 10^{-2} and 10^{-4} Gauß.

There is a corresponding limit from the requirement,[27] that the Larmor motion of the particle fit into the available space; this can be written as $E_{p,max} \simeq 10^{21}\,\mathrm{eV}\,L_{46}^{1/2}$, where L_{46} is the flow of energy along the jet, some fraction of the accretion power to the black hole, in units of 10^{46} erg/s. Therefore radio-galaxies are confirmed source candidates for protons at energies $> 10^{20}$ eV!

4.2. *Positional correlations*

Of course it is interesting to look for associations of ultra high energy events and their arrival directions with known objects in the sky, whether it is the supergalactic plane,[28,29] some distant objects,[30] or nearby galaxies with their black holes. Here we report work with Ioana Maris, done in the years up to 2004, reported on our web-page, and in many lectures.

In order to have a statistical meaningful result we have to form complete samples of candidate sources; in some cases irregular samples will be unavoidable. We need active galactic nuclei, starving and active, starburst and normal galaxies, clusters of galaxies, with the reasoning that all such candidates could produce ultra high energy cosmic rays, be in shocks in jets, in gamma ray bursts, in accretion shocks to clusters, and hyperactive other stars. Recently merged black holes would be another hypothetical option, as then exotic particles might be produced. Most plausible would be relativistic jets pointed at Earth, also known as flat spectrum radio sources; all BL Lac type sources are in this class, although not all flat spectrum radio sources are of BL Lac type.

In any new search, such as to identify sources for high energy neutrinos, or sources for straight propagation of particles in the AUGER data, one should follow the same route: use complete well defined and small samples, and define them beforehand. With the IceCube collaboration we have just completed this task, and it has been published.[31]

We have done this task for ultra high energy cosmic rays some time ago with the data available publicly, using the set of accessible ultra high energy events with good directional accuracy above $4\,10^{19}$ eV: 80 events: AGASA (61), Haverah Park (6), Yakutsk (12), and Fly's Eye (1); we also used one event which was 38 EeV, as this was the sample used also by AGASA for the doublets analysis.

This has been done before by many, from about 1960 by Ginzburg, including us from 1985, and in recent papers.[32–34]

We find, similar to some other searches, that radio sources from the Condon Radio survey in positional coincidence with far infrared (IRAS) sources do show a highly improbable association with ultra high energy cosmic ray events. We can go one step further and use jet-disk symbiosis[35] to predict maximum particle energy and maximum cosmic ray flux, and so check whether these so identified sources can possibly produce the flux observed, even closely. In work with Heino Falcke, Sera Markoff, and Feng Yuan as well as in their own work these concepts about the physics of relativistic jets have been tested at all levels of observability, from microquasars via low luminosity active galactic nuclei to more powerful sources. Current work is being done with Marina Kaufman. Tests include fitting the entire electromagnetic spectrum, and the variability.

Assuming that these particles can violate the GZK interaction with the microwave background, these sources could in fact account for the high flux at high particle energy, quite surprisingly.[36] Also, many of the sources so identified are quite famous in radio astronomy history, sources such as 3C120, 3C147, 4C39.25, 3C449 and the like. We did identify a speculative picture in which such events coming from large cosmological distances could be explained,[37] using particles in higher dimensions and the distortion of space time close to the merger of two spinning black holes. Should be here no such correlation once we have much more extended samples, it might be possible to set limits on this kind of physics.

However, since we tried many times to get such a result we hesitate to assign any physical meaning to this physical picture at this time.

5. Magnetic fields

The magnetic fields filling the cosmos are generally too weak to influence the propagation of ultra high energy cosmic rays by more han a few degrees[38] There may be special environments where this may not be true, like the boundary regions around radio galaxies in a cluster of galaxies.

However, the halos and winds of individual galaxies such as ours may have an appreciable influence: For starburst galaxies such as M82 the existence of a wind has been long shown.[39,40] In our Galaxy this was finally recently demonstrated,[41] for NGC1808 it seems obvious in HST pictures, and the magnetic nature of the wind was demonstrated in the example of NGC4569 through radio polarization images.[42] These last observations also confirm the basic ansatz[43] which we have followed in our work with Laurenţiu Caramete. The key point is that the bending in a magnetic field standard topology such as a Parker wind[44] with $B_\phi \sim \sin\theta / r$ can be large as the Lorentz force is an integral in dr over $1/r$, where r is the distance from the center, and θ is the zenith angle in polar coordinates. Another more subtle point is that the turbulence in the wind is likely to be k^{-2}, a saw-tooth pattern, since it is caused by shock waves running through the medium produced by OB star bubbles and their subsequent supernova explosions. This is in contrast to the turbulence in the thick hot disk,[45] where observations have shown directly, that in a 3D isotropic approximation the turbulence is of Kolmogorov nature, so is $k^{-5/3}$, where k is the wavenumber.

So the general concept of our Galaxy which we use, has three components: A cool thin disk, with lots of neutral and molecular gas, with about 200 pc thickness, but permeated by tunnels of hot gas;[46] a hot disk, with low density and high temperature with a thickness of about 4 kpc, and a magnetic wind, to a very first approximation akin to a Parker wind, extending possibly quite far out, such as a few hundred kpc, or even a Mpc (from ram pressure arguments).

In work with Alex Curuţiu we have calculated in a first approximation the scattering of ultra high energy cosmic rays in such an environment, assuming various possible sources, such as M87 or Cen A. We have tested different assumptions about the level of turbulence, and find that maximal turbulence alone would reproduce the homogeneous sky distribution which has been observed at 30 EeV.

The predicted sky distribution (see the 2004 report on our web-page) consists then of long irregular stripes across the sky, as M87 is quite close to the Galactic North pole. One of these stripes is roughly in the same region as the supergalactic plane in the Southern and part of the Northern sky. This has been our prediction for some years now.

5.1. *Nearby sources*

In other work with Oana Taşcău, Ralph Engel, Heino Falcke, Ralf Ulrich, and Todor Stanev we have identified all available data on nearby black holes, and calculated their cosmic ray maximal contribution, in particle energy and in flux. For most sources the maximal energy of the particles is quite small, below or near EeV, and the flux is also low. However, for a few sources, such as M87, M84, Cen A, and NGC1068 the flux is of interest. At the highest energy only M87 competes, as already suggested many years ago by Ginzburg,[47] and Watson, and in a detailed physical model in Biermann & Strittmatter.[26]

The list has also been available on our web-page since 2004.

There is a question, however, already noted above, that a source may be strong, but how many of these particles make it out of the relativistic bubble (visible in low frequency radio data) around the radio galaxy,[48] and and then on towards us? To fit the data as compiled in the PDG report[49] the flux from M87 has to be reduced by about a factor 4, and so at 1 EeV Cen A is about 20 times stronger than M87, but does not itself extend much beyond 10^{19} eV.

Then just adding the contributions from the strongest sources, and running them through a Monte-Carlo for propagation simulation reproduces the observed spectrum quite well.

5.1.1. *Samples for testing*

Above we have used those active black holes, for which we have data. But other approaches are also possible and would need to be tested. Such sample definitions have recently been developed in collaboration with the IceCube collaboration.[31] Here the sample selection has two key differences: First, ultra high energy cosmic rays are expected to be protons, so they interact with the microwave background, and so almost certainly come from nearby in the cosmos, about 50 Mpc or less. Second, protons are charged and so they may deviate from a straight line in their propagation. Here we focus on the sources.

Even among the nearby sources, there are many possible candidates, one could consider:

- Black holes are quite common, and so almost all galaxies have a massive black hole.[50] Usually the activity of such a black hole is very limited, but experience demonstrates that it is almost always detectable in radio emission, which we interpret as the emission from a relativistic jet.[51] The accretion rate to power this jet could derive from just the wind of a neighboring red giant star. Also, the data suggest that black holes in the centers of galaxies always have a mass larger than about 10^5 or 10^6 solar masses;[52] we have argued that this minimum mass derives from the first growth of a dark matter star,[53,54] possible if in a merger of two galaxies the central dark matter density diverges, builds up a degenerate configuration, a dark matter star, which can then be eaten by a stellar black hole. Dark matter accretion has no Eddington limit, only an angular momentum transport limit, and that should be efficient during the highly disturbed situation of a merger of two galaxies. On the other hand, there might be some subtlety about black hole physics, that we are missing, and so we should obviously just check. The largest challenge to our physical understanding will appear if we find a correlation with very low mass black holes and ultra high energy events. At first, however, this implies just a tally of black holes in our cosmic neighborhood, ordered by mass. And as the mass of the black hole directly relates to the mass of the spheroidal population of stars, or the bulge, as well as the stellar central velocity dispersion, we have to start the search with these properties of galaxies.
- Active black holes, such as M87, NGC315, NGC5128 (= Cen A), all

have relativistic jets, and can accelerate particles. Since the power and also the maximum article energy scale with the radio emission, one needs a sample of black holes with measured compact emission, and so we need a sample ordered by predicted cosmic ray flux at energies beyond about a few EeV. This is what we attempted above.

- If wind supernovae such as exploding Wolf Rayet stars manage to accelerate particles to 3 EeV, which happens to be the cutoff energy also for Galactic confinement, why not even further? These stars do not know about the Galaxy. And if some of these stars under unknown special circumstances produce Gamma Ray Bursts, the particle energy might also be much higher. So this implies a sample of galaxies strong in the far infrared, where strong star formation galaxies, or starburst galaxies, are very prominent. If the flux in cosmic rays scales monotonically with the star formation rate, then a sample of galaxies ordered by flux density at 60 micron would be the best sample to test.
- If the propagation is delayed only slightly in its interaction with the intergalactic magnetic field, then also a direct connection with recent Gamma Ray Bursts might be worth investigating.
- Accretion shocks to clusters of galaxies can also accelerate particles,[55] although probably not to 100 EeV; however, we should actually check with observations. Most clusters, however, are very distant, and such a concept would qualify best for the Virgo cluster, again identifying just one most likely object in the sky, just like the radio galaxy M87, which is one of the dominant galaxies in the Virgo cluster.

6. Predictions

We have presented a theory to account for the entire cosmic ray spectrum beyond the GZK cutoff. This proposal is based on a physical and tested model for relativistic jets.

There is a Galactic magnetic wind, driven by the normal cosmic rays. This wind extends to some fraction of a Mpc. The existence of a Galactic wind in our Galaxy is now confirmed.[41] The basic magnetic field topology is probably of Parker type. The turbulence in the wind is probably sawtooth pattern, i.e. k^{-2}, and its turbulence is close to maximal. All galaxies with an appreciable level of star formation have such a wind, and their environment should look like Swiss cheese, embedded in the supergalactic sheets (work by L. Caramete).

The only contributor for cosmic ray particles beyond the GZK cutoff is M87, with Cen A very strongly contributing just below that characteristic energy, with a small contribution from NGC1068. Weaker sources are negligible due to their low maximum particle energy, and also due to their small flux.

The arrival directions on the sky are smooth around 30 EeV, and begin to become patchy at higher energies, showing some characteristic stripes, in a very simple Parker type model for the magnetic field topology of our Galactic wind.

If the arrival directions are smooth to the highest energies, then this source model fails, and we require Lorentz Invariance Violation,[56] new particles, topological defect or relic decay,[57] dark matter decay or possibly hints of quantum gravity.

Should the proposal be confirmed we can develop sources such 3 C147 as testbeds for particle physics - a CERN / Stanford / Fermilab in the sky.

7. Acknowledgement

P.L. Biermann would like to acknowledge Eun-Joo Ahn, Julia Becker, Venya Berezinsky, Geoff Bicknell, Genadi Bisnovatyi-Kogan, Sabrina Casanova, Mihaela Chirvasa, Alina Donea, Ralph Engel, Torsten Ensslin, Heino Falcke, Paul Frampton, Cristina Galea, Laszlo Gergely, Andreas Gross, Francis Halzen, Alejandra Kandus, Hyesung Kang, Marina Kaufman-Bernardó, Gopal Krishna, Phil Kronberg, Alex Kusenko, Norbert Langer, Hyesook Lee, Sera Markoff, Gustavo Medina-Tanco, Athina Meli, Sergej Moiseenko, Biman Nath, Angela Olinto, Adrian Popescu, Ray Protheroe, Giovanna Pugliese, Wolfgang Rhode, Sorin Roman, Gustavo Romero, Dongsu Ryu, Norma Sanchez, Gerd Schäfer, Eun-Suk Seo, Maury Shapiro, Ramin Sina, Todor Stanev, Jaroslaw Stasielak, Samvel Ter-Antonyan, Valeriu Tudose, Ralf Ulrich, Marek Urbanik, Ana Vasile, Hector de Vega, Yiping Wang, Alan Watson, John Wefel, Stefan Westerhoff, Paul Wiita, Arno Witzel, Gaurang Yodh, Feng Yuan, Cao Zhen, Christian Zier, now T. Kellmann, I. Duţan, L. Caramete, A. Curuţiu, Alina Istrate, ..., ...

Special support comes from the European Union Sokrates / Erasmus grants in collaboration with East-European Universities, with partners T. Zwitter (Ljubljana, Slovenia), L. Gergely (Szeged, Hungary), M. Ostrowski (Cracow, Poland), K. Petrovay (Budapest, Hungary), A. Petrusel (Cluj-Napoca, Romania), and M.V. Rusu (Bucharest, Romania).

Work with PLB is supported through the AUGER theory and membership grant 05 CU 5PD1/2 via DESY/BMBF (Germany), and VIHKOS through the FZ Karlsruhe.

References

1. V.S. Berezinskii, et al., *Astrophysics of Cosmic Rays*, North-Holland, Amsterdam (especially chapter IV) (1990).
2. T. K. Gaisser *Cosmic Rays and Particle Physics*, Cambridge Univ. Press (1990)
3. T. Stanev, *High energy cosmic rays*, Springer-Praxis books in astrophysics and astronomy. Chichester, UK: Springer, 2004.
4. P. L. Biermann, Proc. 23rd International Conference on Cosmic Rays, in Proc. "Invited, Rapporteur and Highlight papers"; Eds. D.A. Leahy et al., World Scientific, Singapore, 1994, p. 45
5. F. Halzen and D. Hooper, *Rep. Prog. Phys.* **65**, 1025 (2002)
6. T. Piran, *Phys. Rep.* **314**, 575 (1999)
7. G. Sigl in "Observing Ultrahigh Energy Cosmic Rays from Space and Earth", Edited by Humberto Salazar, Luis Villaseor and Arnulfo Zepeda, AIP Conference Proceedings, **566**, 266 - 283 (2001)
8. B. Wiebel-Sooth and P. L. Biermann, in Landolt-Börnstein, Handbook of Physics, Springer Publ. Comp., p. 37 - 91, 1999
9. H. Falcke and P. Gorham, *Astropart. Phys.* **19**, 477 (2003)
10. H. Falcke, P. Gorham and R.J. Protheroe *New Astron. Rev.*, **48**, 1487 (2004)
11. H. Falcke et al., LOPES Coll., *Nature* , 435, 313 (2005)
12. W.D. Apel et al.,, *Astropart. Phys.* **26**, 332 (2006) astro-ph/0607495
13. Gemmeke, H., et al., *Inter.Journ. Mod. Phys. A*, **21**, 242 (2006)
14. A. Haungs, LOPES Collaboration, Proceedings of ARENA 06, June 2006, University of Northumbria, UK, (2006a), astro-ph/0610553
15. A. Haungs et al., *Int. Jour. Mod. Phys. A* **21**,182 (2006)
16. J. R. Hörandel et al., *J. of Phys.: Conf. Ser.*, 39, 463 (2006)
17. A. Horneffer et al. *Int. J. of Mod. Phys. A*, **21**, 168 (2006)
18. T. Huege and H. Falcke, *Astropart. Phys.* , 24, 116 (2005)
19. T. Huege, R. Ulrich and R. Engel, *Astropart. Phys.* (submitted), astro-ph/0611742 (2006)
20. S. Nehls et al., *Int. J. of Mod. Phys. A*, **21**, 187 - 191 (2006).
21. J. Petrovic et al., *Jour. of Phys.: Conf. Ser.*, **39**, 471 (2006)
22. S.W. Barwick et al., *Phys. Rev. Lett.* **96**, 171101 (2006)
23. S. W. Barwick et al., 2006, ANITA collaboration, hep-ex/0611008
24. D. Saltzberg et al., *Intern. Journ. of Mod. Phys. A* **21**, 252 (2006)
25. P. Gorham, et al., *Nucl. Instr. and Meth. in Phys. Res. A* **490**, 476 (2002)
26. P.L. Biermann and P.A. Strittmatter, *Astrophys. J.* , **322**, 643 (1987)
27. A.M. Hillas, *Annual Rev. of Astron. & Astrophys.* **22**, 425 (1984).
28. G. de Vaucouleurs, *Vistas in Astron.* **2**, 1584 (1956)
29. T. Stanev, P.L. Biermann, E. Lloyd et al. , *Phys. Rev. Lett.* **75**, 3056 (1995)
30. G.R. Farrar, and P.L. Biermann, *Phys. Rev. Lett.* **81**,3579 (1998)
31. A. Achterberg,: IceCube Collaboration and P. L. Biermann, *Astropart. Phys.* (2006, in press), astro-ph/0609534

32. P. G. Tinyakov, and I.I. Tkachev, *J. of Exp. and Theor. Phys. Lett.*, 74, 445 (2001)
33. N.W. Evans, F. Ferrer and S. Sarkar, *Phys. Rev.* ***D*** **67**, 103005 (2003)
34. Ch.B. Finley, and St. Westerhoff, *Astropart. Phys.* 21, 359 (2004)
35. H. Falcke, and P.L. Biermann *Astrophys. J.* **293** 665 (1995)
36. Biermann, P.L., 2001 - 2006, http://www.mpifr-bonn.mpg.de/div/theory: reports 2001-2004
37. P.L. Biermann and P.H. Frampton, *Phys. Lett. B*, **634**, 125 (2006)
38. D. Ryu, H. Kang, and P. L. Biermann, *Astron. & Astroph.* **335**, 19 (1998)
39. P.P. Kronberg, P.L. Biermann and F. R. Schwab, *Astrophys. J.* **291**, 693 (1985)
40. T.M. Heckman, L. Armus and G.K. Miley, *Astron. J.* **93**, 276 (1987)
41. T. Westmeier, C. Brns and J. Kerp, *Astron. & Astroph.* **432**, 937 (2005)
42. K.T. Chyźy et al. *Astron. & Astroph.* **447**, 465 (2006)
43. D. Breitschwerdt, J. F. McKenzie and H.J. Völk, *Astron. & Astroph.* **245**, 79 (1991)
44. E. N. Parker *Astrophys. J.* **128**, 664 (1958).
45. S. L. Snowden et al., *Astrophys. J.* **485**, 125 (1997)
46. D.P. Cox and B.W. Smith, *Astrophys. J. Letters* **189**, L105 (1974)
47. V.L. Ginzburg and S.I. Syrovatskii, *The origin of cosmic rays*, Pergamon Press, Oxford (1964), Russian edition (1963).
48. F.N, Owen, J.A. Eilek, and N.E. Kassim, *Astrophys. J.* **543**, 611 (2000)
49. T. Gaisser and T. Stanev, T., *Phys. Rev.* ***D*** **66**, id. 010001 (2002)
50. S.M. Faber et al., *Astron. J.* **114**, 1771 (1997)
51. R. Chini, E. Kreysa and P. L. Biermann, *Astron. & Astroph.* **219**, 87 (1989)
52. J.E. Greene, A.J. Barth and L. C. Ho, *New Astron. Rev.* **50**, 739 (2006)
53. F. Munyaneza and P.L. Biermann, *Astron. & Astroph.* **436**,805 (2005)
54. F. Munyaneza and P.L.Biermann, *Astron. & Astroph.* **458**, L9 (2006)
55. H. Kang, J.P. Rachen, and P.L. Biermann, *Month. Not. Roy. Astr. Soc.* **286**, 257 (1997)
56. G. Amelino-Camelia and T. Piran, *Phys. Rev.* ***D*** **64**, 036005 (2001)
57. P. Bhattacharjee and G. Sigl, *Phys. Rep.*, **327**, 109 (2000)

PHYSICS RESULTS OF THE PIERRE AUGER OBSERVATORY

V. VAN ELEWYCK*

for the Pierre Auger Collaboration

Institut de Physique Nucléaire d'Orsay, Université de Paris Sud & CNRS-IN2P3
15, rue G. Clemenceau, 91406 Orsay Cedex, France
** E-mail: vero@ipno.in2p3.fr*

1. Introduction

With three operational fluorescence sites out of four and more than 1000 active Cherenkov detectors on the ground at the time of writing, the Pierre Auger Observatory is nearing completion and has started accumulating data at a regularly increasing pace. In spite of the still small statistics available, a lot of progress has been made in the understanding and fine-tuning of the detector, which has resulted in the development of reliable analysis methods and of the release of its first scientific results concerning the main issues in ultra-high energy (UHE) cosmic ray physics. The spectrum of UHE cosmic rays observed by Auger is presented in [1] as an illustration of the power of Auger's hybrid detection, and the present contribution will focus on the results obtained in the context of anisotropy searches and composition studies.

2. The arrival direction of UHECR: anisotropy studies with the Auger Observatory

Anisotropies in the flux of UHE cosmic rays may appear in different energy ranges and angular scales, depending on the nature, distance and extension of the source. Cosmic rays around an EeV are thought to be of galactic origin, and the region of the Galactic Center and the Galactic Plane are key targets for anisotropy searches performed with Auger data. At higher energies one rather expects UHE cosmic rays to come from extra-galactic sources; a search for directional excesses of cosmic rays could then reveal a correlation with some (un)known astrophysical objects or even exotic sources.

The anisotropy studies performed by Auger are based on all surface detector (SD) events (plus some hybrids) with zenith angle $\theta < 60°$ that pass the quality cut T5, which requires that the station with the highest signal be surrounded by a hexagon of working stations, ensuring a good reconstruction of the event. The energy of the events is determined using the constant intensity cut method and calibrating the S_{38} parameter to the energy obtained from the florescence detector (FD) as described in [1]. .

2.1. *Angular resolution and coverage maps*

To detect an excess of events coming from a particular region of the sky, one has to compare the number of events observed in that region with the corresponding coverage map, that is, the background number of events expected from an isotropic flux of cosmic rays in the same exposure conditions. This procedure requires accurate knowledge of the detector properties, and in particular of its angular accuracy and of the exposure dependance in time, energy and solid angle for each point of the sky. For a more detailed discussion of these issues, we refer the reader to [2,3].

The angular resolution AR for the SD is defined as the angular radius that would contain 68% of the showers coming from a point source; it is determined from the zenith (θ) and azimuth (ϕ) uncertainties obtained from the geometrical reconstruction on an event by event basis,

$$AR = 1.5\ \sqrt{[\sigma^2(\theta) + sin^2(\theta)\,\sigma^2(\phi)]\,/2} \tag{1}$$

where σ^2 is for the variance. The AR is driven by the accuracy on the measurement of the arrival time of the shower front in each station. The variance on the arrival time T_1 of the first particle is parameterized according to the time variance model described in [2,3], which was validated using data from the so-called "doublets" (pairs of tanks separated by 11 m).

To build the coverage maps, one has to consider all possible modulations and inhomogeneities in the exposure of different regions of the sky. Besides the obvious effects due to the rotation of the Earth and the limited field of view of the detector, other modulations are induced by the continuously growing size of the array, by temporary failures in some detectors, and by temperature and pressure variations (which affect both the shower development in the atmosphere and the response of the electronics). Two different techniques have been used to estimate the SD coverage maps [4]:

- **the semi-analytic method** consists in an analytical fit to the θ distribution of the events in the relevant energy range, convoluted with an acceptance factor which accounts for the time evolution of the detector (according to the trigger activity), assuming a uniform response in azimuth (which is valid for showers up to 60°).

- **the shuffling technique** takes the average of many fake data sets generated by shuffling the observed events in such a way that the arrival times are exchanged and the azimuths are drawn uniformly. This shuffling also preserves the θ distribution of the events. It might partially absorb an intrinsic large-scale anisotropy present in the cosmic ray flux, but this drawback can be avoided using independent shufflings in (day x hour).

The expected number of events in a given pixel of the sky is obtained by integrating the coverage map in a given window, while the signal is determined by applying the same filtering to the event map. A significance map is then generated by comparing the signal in each pixel respect to the expected background, according to the Li & Ma procedure [5].

2.2. *Anisotropy studies around the galactic center*

The region of the Galactic Center (GC, located at the equatorial coordinates $(\alpha, \delta) = (266.3°, -29.0°)$) and the Galactic Plane (GP) are particularly attractive targets for cosmic ray anisotropy studies around EeV energies. Two cosmic ray experiments, AGASA and SUGAR, have already claimed significant excesses in the flux of UHECR in that region. AGASA [6] reported a 4.5σ excess of CR with energies in the range $10^{18} - 10^{18.4}$ eV in a 20° radius region centered at $(\alpha, \delta) \simeq (280°, -17°)$ (it is worth noting however that the GC itself lies outside of the AGASA field of view). Subsequent searches near this region using old SUGAR data [7] failed to confirm that result but found a 2.9σ excess flux of CR in the energy range $10^{17.9} - 10^{18.5}$ eV in a 5.5° window centered at $(\alpha, \delta) \simeq (274°, -22°)$. Recent observations by HESS of a TeV γ ray source in that region [8] and of diffuse γ-ray emission from the central 200 pc of the GP [9] have provided additional hints towards the presence of powerful CR accelerators in the Galaxy. In that context, several models that predict a detectable flux of neutrons in the EeV range (whose decay length is about the distance from the GC to the Earth) have also been proposed.

With the GC well in the field of view and an angular resolution which is much better than previous CR experiments, the Pierre Auger Observatory is well suited to look for UHECR anisotropies coming from that region. A total of 79265 SD events and 3934 hybrid events have been used, which cor-

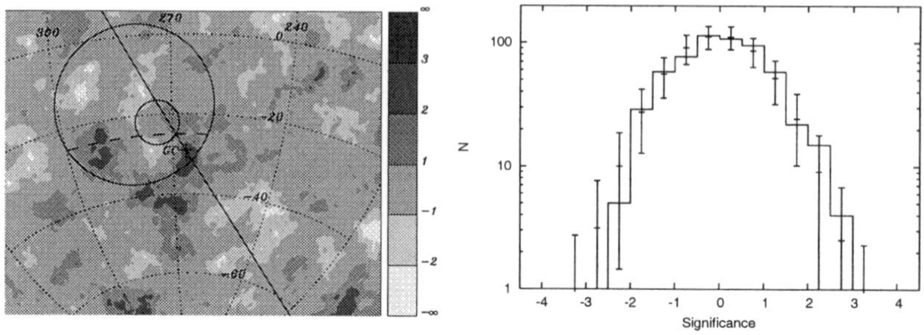

Fig. 1. *Left:* significance map of CR overdensities in the region of the Galactic Center in the energy range $10^{17.9} - 10^{18.5}$ eV, showing the Galactic Center (cross), the Galactic Plane (solid line), the regions of excesses of AGASA and SUGAR (circles), and the AGASA field of view limit (dashed line). The event map was smoothed with a top-hat 5.5° window. *Right:* corresponding histogram of overdensities computed on a grid of 3° spacing, compared to the average isotropic expectations points (with 2σ bars). (from [10])

responds to the data collected between January 2004 and March 2006 satisfying the T5 quality cut [1] and with $\theta < 60°, 10^{17.9}$ **eV** $< E < 10^{18.5}$ **eV**; it represents respectively more than four and ten times the sample that AGASA and SUGAR used in this context.

Significance maps were built using different filterings of the data to account for the angular size of the excesses reported by AGASA and SUGAR in their respective energy range. An example of such a map is shown in Fig. 1 together with the corresponding overdensity distribution. Several tests were also performed with modified energy windows to account for a possible energy shift due to differences in the calibration of the experiments. In all cases, Auger data have been found compatible with isotropy, therefore not confirming the results from previous experiments. Even in the worst case of a source emitting nucleons and embedded in a background made of heavier nuclei, to which Auger is more sensitive in the relevant energy range, a significant excess (5.2σ) would be expected, in contradiction with current observations. Details of the analysis can be found in [10].

Data from the Auger Observatory were also used to search for a point source in the direction of the GC itself at the scale of Auger's own angular resolution. In the energy range $10^{17.9}$ – $10^{18.5}$ eV, and applying a 1.5° Gaussian filter to account for the pointing accuracy of the SD, we obtain 53.8 observed events against 45.8 expected. This allows to put a 95% C.L. upper bound on the number of events coming from the source of $n_s^{95} = 18.5$. Assuming that, in the energy range considered, both the source and the bulk CR spectrum have similar spectral indexes and that the emitted CR are proton-like, and taking a differential spectrum $\Phi_{CR}(E) \simeq \xi\ 30\ (E/EeV)^{-3}\ EeV^{-1}km^{-2}yr^{-1}sr^{-1}$, where ξ parameterizes the uncertainties on the flux normalization, a 95% C.L. upper bound of

$$\Phi_s^{95} \leq \xi\ 0.08\ km^{-2}yr^{-1} \tag{2}$$

can be set on the source flux. This bound could however be about 30% higher if the CR composition at EeV were heavy, *i.e.* close to Iron.

Finally, a scan for correlations of CR arrival directions with the Galactic Plane and Super-Galactic Plane have also been made in two different windows of energy (1 EeV < E < 5 EeV and E > 5 EeV), yielding again negative results (although with a smaller dataset) [11].

2.3. *Other searches for localized excesses in the Auger sky maps*

The Auger data have also been used to perform both blind searches and prescripted searches for localized excesses in other parts of the sky.

In the case of blind searches, the distribution of significances is compared to those obtained from a large number of Monte Carlo isotropic simulations. Such searches were performed both for a 5° and a 15° angular scale and in two separate energy ranges, $1\mathbf{EeV} \leq E \leq 5\mathbf{EeV}$ and $E \geq 5\mathbf{EeV}$; all of them turned out to be compatible with isotropy [12].

The Auger Collaboration had also released a list of prescribed targets with definite angular and energy windows [13], with the associated significance probability level to attain in order to claim a positive signal. The prescription targets range from the Galactic Center to some nearby violent extragalactic objects; none of them has turned out to lead to a positive detection. As more data is streaming in, the catalogue of candidate targets that will be studied is expected to increase in the future.

3. The nature of UHECR: composition studies with the Auger Observatory

Thanks to its hybrid capabilities, the Auger Observatory can extract complementary information on the shower development parameters, that are ultimately related to the nature of the primary cosmic rays. If the discrimination between different types of nuclei is complicated by the uncertainties in the hadronic models governing the interactions of the particles in the

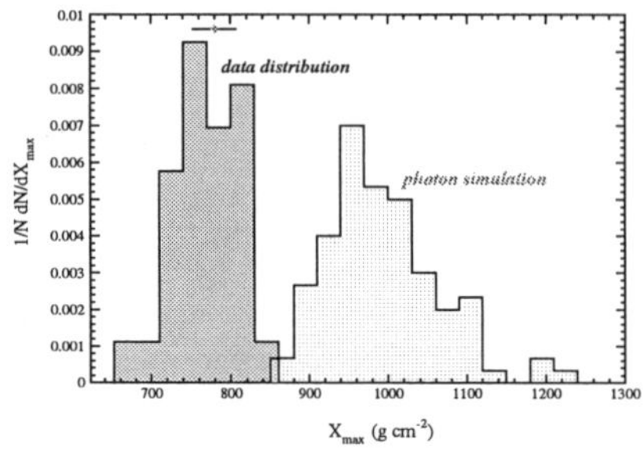

Fig. 2. X_{max} distribution of simulated and real candidate photon events. The blue point is the X_{max} value and uncertainty for one event from the data.

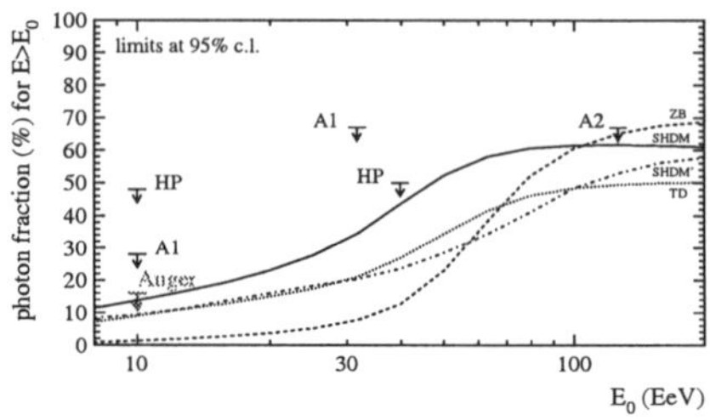

Fig. 3. 95% C.L upper limit on the photon fraction in UHE cosmic rays obtained by Auger, compared to the results of Haverah Park [15], HP, and AGASA [16], A1 and A2.

shower at such high energies, several methods have already been proposed for the identification of photons and neutrinos, and are currently applied to the data of the Auger Observatory. The presence and the amount of photons and neutrinos at such high energies would constitute a crucial probe for many exotic models of UHE cosmic ray production and could help locate candidate sources as they travel undeflected by the intergalactic magnetic fields.

3.1. *Upper limit on the UHE photon flux*

Unlike protons and nuclei, the development of photon showers are driven by electromagnetic (EM) interactions and do not suffer much from the uncertainties in hadronic interactions. Photon showers are expected to contain fewer and less energetic secondary muons, as a result of the smallness of the photon radiation length respect to its mean free path for photo-nuclear interactions and direct muon production. Their development is also delayed due to the small multiplicity in EM interactions and to the LPM effect [14], which reduces the bremsstrahlung and pair production cross-sections at energies above 10 EeV. These considerations allowed several ground array experiments to set upper limits on the flux of UHE photons on basis of studies of the rate of vertical to inclined showers (in Haverah Park experiment [15]) and of the muon content of the showers at ground (in AGASA [16]).

Taking profit of its hybrid design, Auger has set up a different method to identify photon primaries in the flux of UHECR. It is based on the direct observation of the longitudinal profile of the shower development in the atmosphere by the FD, and uses as a discriminating variable the atmospheric depth of the shower maximum, X_{max} (the estimated average difference in X_{max} between photons and hadrons is about 200gr/cm^2).

The data set used for this analysis corresponds to the hybrid events (*i.e.* those observed by one or more FD telescope and by at least one SD station, which ensures a better angular accuracy and smaller uncertainty in the reconstruction of X_{max}) with a reconstructed energy $E > 10^{19}$ eV, registered between January 2004 and February 2006. During that period two of the four Auger eyes were active (for a total of 12 FD telescopes) and the number of deployed SD stations grew from 150 to 950.

A series of cuts were applied to the data that guarantee the quality of the hybrid geometry and of the fit to the shower longitudinal profile, which takes into account the local amtospheric conditions (see the detail in [17]). One important condition is to have the X_{max} of the shower inside the field of view of the telescopes. To minimize the bias that this condition introduces against photon primaries in the detector acceptance, additional energy-dependant cuts are applied both on the zenith angle and the maximum distance of telescope to shower impact point in order to eliminate nearly-vertical and distant events.

For each of the 29 events that survived all the cuts, 100 photon showers were simulated in the same energy and arrival direction conditions and the resulting expected distribution of X_{max} was compared to the observed X_{max} of the event. An example is shown in Fig. 2, together with the distribution of the X_{max} from the whole selected dataset. For all 29 events, the observed X_{max} is well below the average value expected for photons.Taking systematic uncertainties on the X_{max} determination and the photon shower simulations into account, the available statistics allows to put an upper limit on the photon fraction of 16% at 95% C.L, which is shown in Fig. 3 together with previous results and some predictions from non-accelerator models.

3.2. *Inclined air showers and the detection of neutrinos*

The use of Cherenkov water tanks for the SD allows the Pierre Auger Observatory to detect showers with zenith angles up to 90° (and even more) [18]. The range of inclined showers, $60° \leq \theta \leq 90°$, contributes half the total solid angle of the detector and about 25% of its geometrical acceptance, thereby significantly increasing the field of view of the detector and the SD statistics. Such events are indeed seen by Auger both in the SD and the FD; some of them may be quite spectacular, with very extended footprints involving tens of tanks, as illustrated in Fig. 4. Dedicated selection pro-

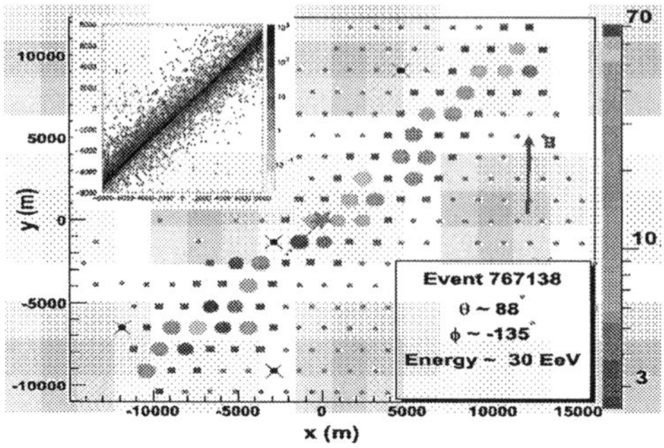

Fig. 4. Example of a near horizontal air shower as seen by the SD; the shower triggered 31 tanks and extends on about 30 km at ground. In the upper left corner, the best fitting simulated muon map corresponding to the reconstructed zenith and azimuth angles.

cedures and reconstruction methods are being developed in Auger to deal with the distinctive features of those showers.

The distance between the first interaction point (normally in the first few 100 g cm^{-2}) and the detector position is much larger than in the vertical case, the atmospheric depth ranging from 1740 g cm^{-2} at 60° till 31 000 g cm^{-2} at 90°). As a result, the EM component of the shower dies out long before reaching the ground, and the only particles recorded in the SD are energetic muons (typically of 10-1000 GeV) accompanied by an EM halo which is constantly regenerated by muon decay, brehmsstrahlung and pair production. Those muons arrive at ground in a thin front with small curvature, resulting in short FADC pulses in the tanks, as shown in Fig.5 (right). Their trajectories are long enough to be affected by the geomagnetic field, which leads to a separation between positive and negative muons and a further elongation of the projected footprint on the ground.

The reconstruction of inclined showers is based on the search for the best fit to the pattern of signals at ground performed with averaged maps of muon densities obtained from simulations. The relation between the muon density and the energy depends on the nature of the primary cosmic ray, and is established on basis of Monte Carlo simulations which suffer from hadronic interactions uncertainties at high energies. In this context, hybrid inclined events reconstructed by both the SD and the FD will play an important rôle in primary composition studies, since they allow independant measurements of the EM and muonic components of the shower [19].

Inclined showers also constitute the bulk of events from which a signal of UHE cosmic neutrino could be extracted. Due to their small cross-section,

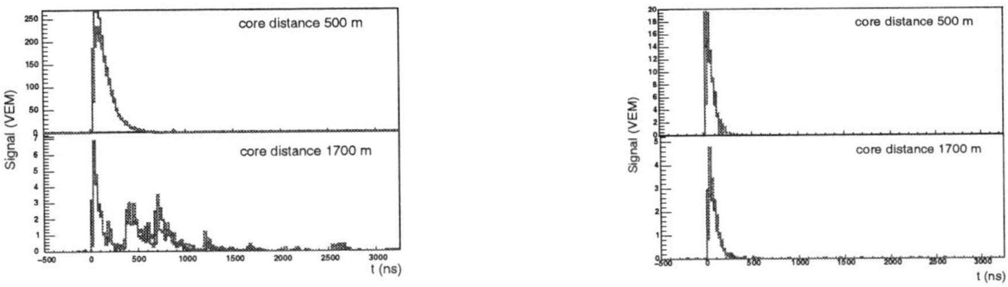

Fig. 5. FADC traces of a young (*left*) and old (*right*) shower. The signal from a young shower gets smaller and more extended as the distance to the core increases, while old showers have short traces at all distances.

neutrinos can penetrate deeply in the atmosphere and initiate showers at all possible depths, unlike nuclei or photons. In particular, showers originating less than ≈ 2000 g cm^{-2} away from the detector will reach it before their EM component attenuates completely. Selection criteria will thus require the presence of signals corresponding to a young shower, and in particular of stations with extended traces that reflect the large curvature of the shower front and the presence of an EM component (see Fig. 5).

Up-going tau neutrinos that skim the Earth just below the horizon could also be detected as they are likely to interact in the ground and produce a tau which may emerge from Earth and initiate an observable air shower, provided it decays close enough to the SD. Preliminary studies provided a proof of principle for the detection of such neutrinos in the energy range $10^{17} - -10^{19}$ eV [20] and, although a careful study of systematic uncertainties is necessary to infer with a reasonable precision the energy of the incident ν_τ primary, this method seems the most promising in terms of acceptance, which is a crucial matter when dealing with event rates as small as ~ 1 per year. Studies are currently ongoing both in the down-going and upgoing ranges to define and optimize the selection criteria, and the search for UHE neutrinos in the Auger data has started.

4. Conclusions

The Southern Auger Observatory, expected to be complete in 2007, has delivered its first science results on the UHE cosmic ray spectrum, anisotropy searches and composition studies. In particular, the region of the Galactic Center has been studied with a precision never attained before, yielding no hint of anisotropies. The absence of evidence for a point-source near the GC excludes several scenarios of neutron sources recently proposed. The upper limit on the photon fraction above 10 EeV, derived for the first time from a direct observation of the shower maximum, confirms and improves previous limits from ground arrays. Finally, the inclined shower data sample will soon contribute to enlarge the field of view of the detector and increase its statistics; it might also reveal the first cosmic neutrino ever observed at ultra-high energies.

Acknowledgements

Many warm thanks are due to the directors and organizing staff of the School, profs. M. Shapiro, T. Stanev, J. Wefel and A. Smith, for generating a lively and inspiring scientific atmosphere in Erice. I am grateful to prof. J. Cronin for giving me the opportunity to present the Auger results to such a rewarding audience. This work was supported by the European Community 6th F.P. through the Marie Curie Fellowship MEIF-CT-2005 025057.

References

1. P. Privitera, *The Auger Observatory*, these Proceedings.
2. C. Bonifazi [Pierre Auger Collaboration], Proc. 29th ICRC **7** (2005) 17.
3. A. Letessier-Selvon [Pierre Auger Collaboration], arXiv:astro-ph/0610160.
4. J.-Ch. Hamilton [Pierre Auger Collaboration], Proc. 29th ICRC **7** (2005) 63.
5. T.-P. Li and Y.-Q. Ma, Astrophys. J **272** (1983) 317.
6. N. Hayashida *et al.* [AGASA Collaboration], Astropart. Phys. **10** (1999) 303 ; M. Teshima *et al.* [AGASA Collaboration], in Proc. 27th ICRC **1** (2001) 337.
7. J. A. Bellido *et al.*, Astropart. Phys. **15** (2001) 167.
8. F. Aharonian *et al.* [HESS Collaboration], Astron. Astrophys. **425** (2004) L13.
9. F. Aharonian *et al.* [HESS Collaboration], Nature **439** (2006) 695.
10. M. Aglietta *et al.* [Pierre Auger Collaboration], Astropart. Phys., in press [arXiv:astro-ph/0607382].
11. A. Letessier-Selvon [Pierre Auger Collaboration] Proc.29th ICRC **7**(2005) 67.
12. B. Revenu [Pierre Auger Collaboration], Proc. 29th ICRC **7** (2005) 75.
13. R. Clay [Pierre Auger Collaboration], Proc. 28th ICRC **1** (2003), 421.
14. L. D. Landau, I. Ya. Pomeranchuk Dokl. Akad. Nausk. SSSR **92** (1953), 535 & 735; A. B. Migdal, Phys. Rev. **103** (1956), 1811.
15. M. Ave *et al.*, Phys. Rev. Lett. **85** (2000), 2244; Phys. Rev. **D65** (2002) 063007.
16. K. Shinozaki *et al.*, Astrophys. J. **571** (2002), L117; M. Risse *et al.*, Phys. Rev. Lett. **95** (2005),171102.
17. J. Abraham *et al.* [Pierre Auger Collaboration], Astropart. Phys., in press [arXiv:astro-ph/0606619].
18. L. Nellen [Pierre Auger Collaboration], Proc. 29th ICRC **7** (2005), 183; V. Van Elewyck [Pierre Auger Collaboration], AIP Conf. Proc. **819** (2006), 187.
19. M. Ave *et al.*, Proc. 28th ICRC **1** (2003), 563.
20. K.S. Capelle *et al.*, Astropart. Phys. **8** (1998), 321; X. Bertou *et al.*, Astropart. Phys. **17** (2002), 183.

THE KASCADE–GRANDE EXPERIMENT

F. Cossavella[a] *, W.D. Apel[b], J.C. Arteaga[b,1], F. Badea[b,2], K. Bekk[b], A. Bercuci[c], M. Bertaina[d], J. Blümer[b,a], H. Bozdog[b], I.M. Brancus[c], M. Brüggemann[e], P. Buchholz[e], A. Chiavassa[d], K. Daumiller[b], F. Di Pierro[d], P. Doll[b], R. Engel[b], J. Engler[b], P.L. Ghia[f], H.J. Gils[b], R. Glasstetter[g], C. Grupen[e], A. Haungs[b], D. Heck[b], J.R. Hörandel[a], T. Huege[b], P.G. Isar[b,3], K.-H. Kampert[g], H.O. Klages[b], Y. Kolotaev[e], P. Luczak[h], H.J. Mathes[b], H.J. Mayer[b], C. Meurer[b], J. Milke[b], B. Mitrica[c], C. Morello[f], G. Navarra[d], S. Nehls[b], R. Obenland[b], J. Oehlschläger[b], S. Ostapchenko[b,4], S. Over[e], M. Petcu[c], T. Pierog[b], S. Plewnia[b], H. Rebel[b], A. Risse[h], M. Roth[b], H. Schieler[b], O. Sima[c], M. Stümpert[a], G. Toma[c], G.C. Trinchero[f], H. Ulrich[b], J. van Buren[b], W. Walkowiak[e], A. Weindl[b], J. Wochele[b], J. Zabierowski[h], D. Zimmermann[e]

[a] *Institut für Experimentelle Kernphysik, Universität Karlsruhe, 76021 Karlsruhe, Germany,*
[b] *Institut für Kernphysik, Forschungszentrum Karlsruhe, 76021 Karlsruhe, Germany*
[c] *National Institute of Physics and Nuclear Engineering, 7690 Bucharest, Romania*
[d] *Dipartimento di Fisica Generale dell'Università, 10125 Torino, Italy*
[e] *Fachbereich Physik, Universität Siegen, 57068 Siegen, Germany*
[f] *Istituto di Fisica dello Spazio Interplanetario, INAF, 10133 Torino, Italy*
[g] *Fachbereich Physik, Universität Wuppertal, 42097 Wuppertal, Germany*
[h] *Soltan Institute for Nuclear Studies, 90950 Lodz, Poland*

[1] *permanent address: CINVESTAV, Mexico D.F., Mexico*
[2] *on leave of absence from* [c]
[3] *on leave of absence from Nat. Inst. Space Science, Bucharest, Romania*
[4] *on leave of absence from Moscow State University, 119899 Moscow, Russia*

* *E-mail: fabiana.cossavella@ik.fzk.de*

The KASCADE–Grande experiment measures extensive air showers induced by primary cosmic rays in the energy range of $10^{14} - 10^{18}$ eV. As extension of the original KASCADE experiment it allows the investigation of the knee and the possible second knee in the cosmic ray energy spectrum. An overview of the experimental setup and preliminary results are given.

Keywords: KASCADE–Grande; EAS; cosmic rays

1. Introduction

The extensive air shower experiment KASCADE [1] has shown that at energies of a few times 10^{15} eV the knee in the cosmic ray energy spectrum is due to light elements and that its position depends on the kind of incoming particles, suggesting a possible rigidity dependence. At an energy of approximately $3-7\cdot 10^{17}\,\mathrm{eV}$ some experiments [2–4] report a steepening in the spectrum, usually referred to as the "second knee". According to some astrophysical scenarios, like the change of acceleration mechanisms at cosmic ray sources, the knee of the heavy component of cosmic rays is expected at a primary energy of around $Z_{Fe}\cdot E_{1knee}\approx 10^{17}\,\mathrm{eV}$. Another possible origin of the second knee could be the transition from galactic to extragalactic cosmic rays.

Due to the low flux of cosmic rays in the order of $10^{-10}\,\mathrm{m}^{-2}\,\mathrm{s}^{-1}\,\mathrm{sr}^{-1}$ for energies above 10^{17} eV, the collective area of KASCADE is not sufficient for investigations in this energy range.

KASCADE-Grande is, thus, the natural extension of KASCADE over an area of approximately $0.5\,\mathrm{km}^2$, suitable for detection of primary particles up to energies of 10^{18} eV. Its main goals are the investigation of the possible existence of the iron knee and the nature of the second knee.

2. Experimental setup

KASCADE-Grande is located at the Forschungszentrum Karlsruhe, Germany, at 110m above sea level. The field array of the original KASCADE [5] experiment consists of 252 stations placed on a grid of $200\times 200\,\mathrm{m}^2$. Each station houses liquid scintillators for the detection of e/γ and shielded plastic scintillators for the muonic component, with a total coverage of $490\,\mathrm{m}^2$ for e/γ ($E_e > 5\,\mathrm{MeV}$) and $622\,\mathrm{m}^2$ for μ ($E_\mu > 230\,\mathrm{MeV}$). A muon tracking detector, with 3 horizontal layers of streamer tubes of $128\,\mathrm{m}^2$ each and 2 vertical layers on both sides, measures and tracks the single muons with an energy threshold of 800 MeV. Muons with $E^{th}=2.4\,\mathrm{GeV}$ are measured by multiwire/proportional chambers and limited streamers tubes, placed in the Central Detector over an area of $300\,\mathrm{m}^2$.

Grande extends KASCADE by an array of 37 detector stations, organized into 18 hexagonal trigger cells of 7 stations each (Fig. 1). Each station consists of 16 scintillation detectors ($80\times 80\times 4\,\mathrm{cm}^3$), arranged in a 4×4 grid, for a total surface of $10\,\mathrm{m}^2$. Each scintillator is read out by a high gain photomultiplier. The four central scintillators are simultaneously read out by one low gain PMT each, to cover a dynamic range from 0.3 to 6000 particles/m^2. Full efficiency is reached with a 7 out of 7 stations coincidence (0.5 Hz) at a primary energy of $\approx 2\times 10^{16}$ eV, as shown in Fig. 1. In order to provide a fast trigger to the KASCADE muon tracking and central detector set-ups, there exists the Piccolo cluster, comprising 8 stations of

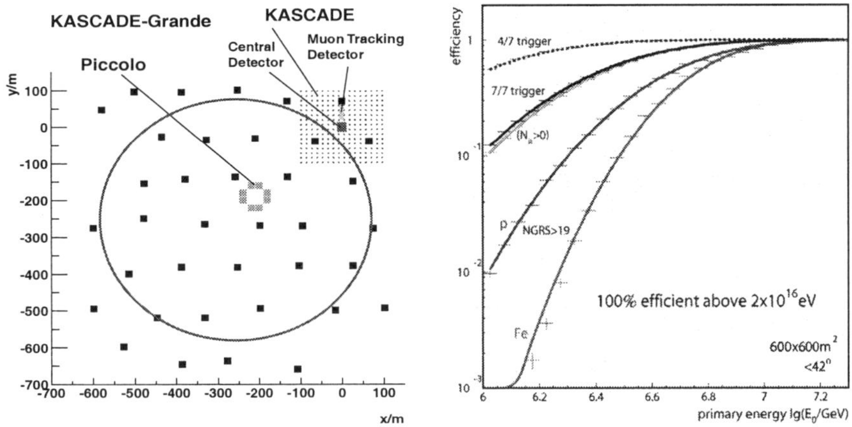

Fig. 1. Left: the Grande array with 37 detector stations, the Piccolo cluster, KASCADE array and Central Detector.The circle describes the area of reconstruction with high accuracy. Right: trigger and reconstruction efficiency for showers between 10^{15} and 10^{18} eV, with a zenith angle smaller than 42°. Requiring a 7/7 trigger at Grande with muon number successfully reconstructed by KASCADE, a full efficiency is reached at 2×10^{16} eV. Common data quality cuts require at least 19 active Grande detector stations, for which case the efficiency is plotted for proton and iron.

plastic scintillators, located between the KASCADE array and the center of Grande.

In addition there are 30 dipole radio antennas, mainly spread over the area of the KASCADE array, forming the LOPES experiment [6] for the measurement of the radio emission from air showers.

3. Reconstruction and accuracy

Analysis of Grande array data provides information on core position, arrival direction and the total number of charged particles (N_{ch}) in the shower. The muon number N_μ is retrieved from the KASCADE array data. By subtracting N_μ from N_{ch} it is possible to calculate the electron shower size N_e.

The lateral distribution of electrons has been studied through detailed CORSIKA [7] simulations and is described best in case of KASCADE-Grande by a modified NKG-function [8]:

$$\rho_e = N_e \cdot C(s) \cdot \left(\frac{r}{r_0}\right)^{s-\alpha} \cdot \left(1 + \frac{r}{r_0}\right)^{s-\beta}$$

where the normalization factor is $C(s) = \Gamma(\beta - s)/(2\pi r_0^2 \Gamma(s - \alpha + 2)\Gamma(\alpha + \beta - 2s - 2))$ and s is the shower age. From simulations, values of $\alpha = 1.5$,

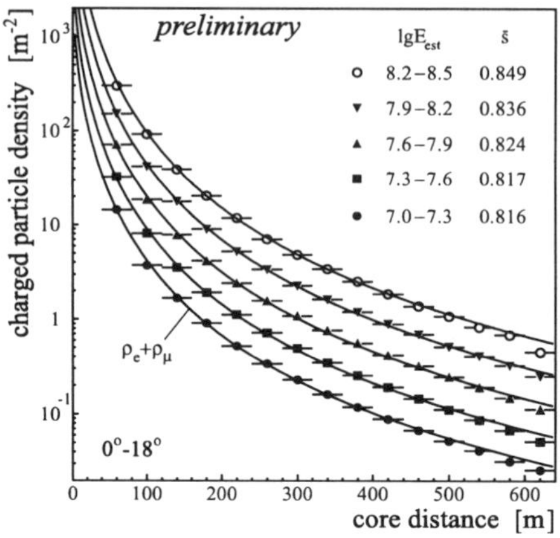

Fig. 2. Measured lateral distribution of charged particles for 5 different energy bins above 10^{17} eV [8]

$\beta = 3.6$ and $r_0 = 40\,\mathrm{m}$ were found as optimum for the radial distances relevant for Grande.

For the lateral distribution of muons, a modified Lagutin function [9] :

$$\rho_\mu = N_\mu \, \frac{0.28}{r_0^2} \left(\frac{r}{r_0}\right)^{-0.69} \cdot \left(1 + \frac{r}{r_0}\right)^{-2.39} \cdot \left[1 + \left(\frac{r}{10 \cdot r_0}\right)^2\right]^{-1}$$

is used, with $r_0 = 320\,\mathrm{m}$. The good agreement of lateral distribution functions with real data is shown in Fig. 2, where, for different primary energy bins, fit functions $\rho_e + \rho_\mu$ with average fit parameters are superimposed on the data.

To test the reconstruction procedure and estimate the uncertainty, showers generated by CORSIKA have been used as input for a detailed GEANT [10] simulation of the apparatus. Approx. 200,000 proton and iron showers in the energy range of $10^{15} - 10^{18}$ eV, with zenith angles between 0° and 18°, have been analyzed with the same procedure as used for real data. The results for spatial and directional resolution are shown as a function of the shower size in Fig. 3: above the threshold of 10^6 electrons (corresponding to 100% trigger efficiency) the resolution is better than 12 m and 0.6°. Fig. 3 also displays the accuracy of the muon and electron numbers. The statistical uncertainty, expressed by the error bars, is around 25% at threshold and decreases with increasing shower size as expected, while the systematic deviation (average difference between reconstructed and true logarithmic value) decreases from 0 to -0.05. An analogous plot for the es-

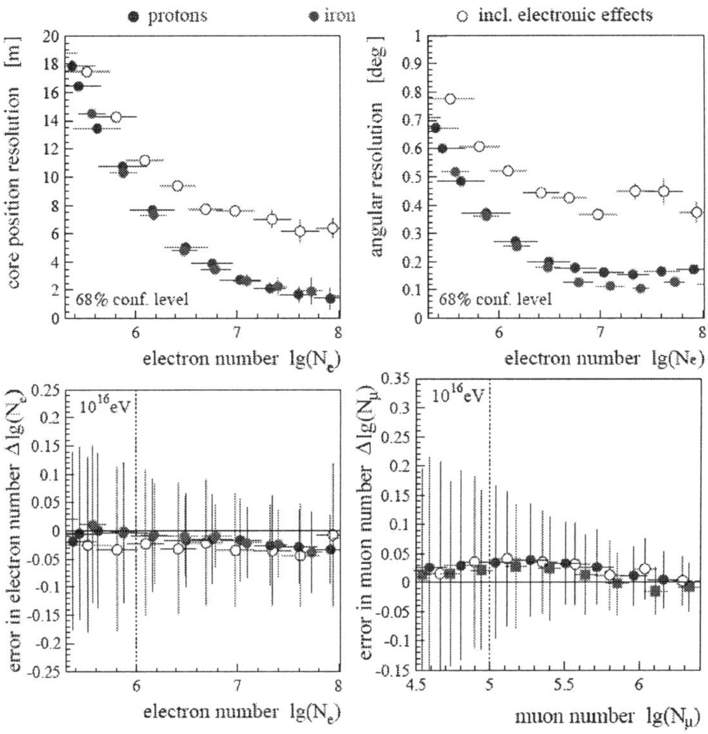

Fig. 3. Reconstruction accuracies for core position (upper left), shower direction (upper right) as a function of electron number. On the bottom: systematic deviation of reconstructed electron and muon numbers. The error bars indicate statistical errors of single reconstruction [8]

timation of the muon number shows a small systematic overestimation of the muon component.

Comparison of real events reconstructed independently by Grande and KASCADE confirms the values we obtained for the uncertainty in core and angular resolution, with an error of 10 m for core position and 0.8° for arrival direction at threshold.

4. First results

With the capability of reconstructing both, muon and electron numbers, it is possible to investigate a two-dimensional size spectrum. The present data set for zenith angles below 18° is shown in Fig. 4, dashed lines show

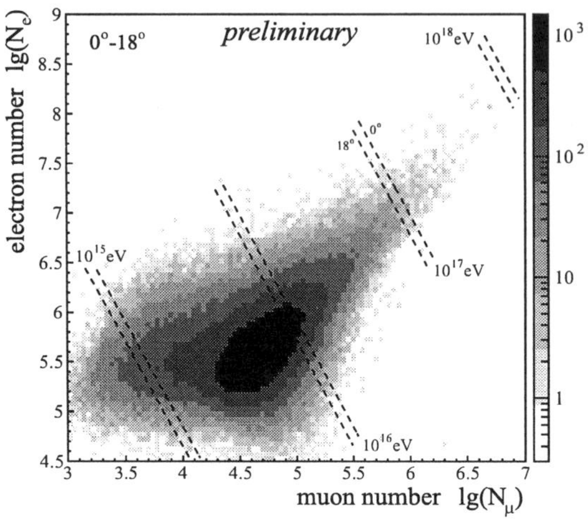

Fig. 4. Reconstructed electron and muon number distribution of air showers measured by KASCADE-Grande. The dashed lines indicate average lines of constant energy derived from CORSIKA simulations [8]

an estimation of the primary energy based on simulations with the interaction model QGSjet01. With one year of effective data taking, KASCADE-Grande has collected the same statistics in the overlapping energy region $10^{16} - 10^{17}$ eV as KASCADE did in ten years. At the moment the statistics are too small to make a concise statement of the spectra of mass groups at energies above 10^{17} eV. This spectrum will be the starting point for the application of an unfolding analysis that will lead to the determination of spectra for different mass groups (as done for KASCADE [1]).

References

1. T.Antoni et al. - KASCADE Collaboration, *Astrop. Phys.* 24 (2005), p. 1-25
2. T.Abu-Zayyad et al., *Astrophys. J.* 557 (2001), p. 686-699
3. D.J.Bird et al., *Astrophys. J.* 424 (1994), p. 491-502
4. V.P. Egorova et al., *Nucl. Phys. B (Proc. Suppl.)* 136 (2004), p. 3
5. T.Antoni et al - KASCADE Collaboration, *Nucl. Instr. Meth.* A 513 (2003)
6. H.Falcke et al. - LOPES Collaboration, *Nature* 435 (2005), p. 313-316
7. D.Heck et al., *Report FZKA 6019*, Forschungszentrum Karlsruhe (1998)
8. R.Glasstetter et al.- KASCADE-Grande Coll., *Proc. of* 29th *ICRC* 6 (2005), p. 293-296
9. J.van Buren et al.- KASCADE-Grande Coll, *Proc. of* 29th *ICRC* 6 (2005), p. 301-304
10. *GEANT - Detector Desc. and Sim. Tool*, CERN Program Library Long Writeup, W5013, CERN (1993)

MEASUREMENT OF THE RELATIVE ABUNDANCES OF THE ULTRA-HEAVY GALACTIC COSMIC-RAY ABUNDANCES ($30 \leq Z \leq 40$) WITH TIGER

B.F. Rauch[a*], L.M. Barbier[b], W.R. Binns[a], J.R. Cummings[b], G.A. de Nolfo[b], S. Geier[c] M.H. Israel[a], J.T. Link[a], R.A. Mewaldt[c], J.W. Mitchell[b], S.M. Schindler[c], L.M. Scott[a], E.C. Stone[c], R.E. Streitmatter[b] and C.J. Waddington[d]

(a) Washington University, St. Louis, MO 63130, USA
(b) Goddard Space Flight Center, Code 661, Greenbelt, MD 20771, USA
(c) California Institute of Technology, Pasadena, CA 91125, USA
(d) University of Minnesota, Minneapolis, MN 55455, USA
** E-mail: brian@cosray.wustl.edu*

Observations of Ultra-Heavy galactic cosmic rays (GCR) help to distinguish the possible origins of GCRs. The Trans-Iron Galactic Recorder (TIGER) is designed to measure the charge (Z) and energy of GCRs using a combination of scintillation counters, Cherenkov counters, and a scintillating fiber hodoscope. TIGER has accumulated data on two successful flights from McMurdo, Antarctica: the first launched in December of 2001 with a total flight duration of 31.8 days and the second in December of 2003 with a total flight duration of 18 days. The two flights of TIGER achieved sufficient statistics and charge resolution to resolve ~140 particles with $Z > 30$, and have provided the best measurements to date for Zn, Ga, Ge, and Se. We present a preliminary analysis of the combined data from both flights for Ultra-Heavy GCRs and discuss the results in the context of different GCR source models.

Keywords: Galactic cosmic rays; Galactic abundances.

1. Introduction

The principal objective of TIGER is the determination of the source abundances of the heavy GCRs with $Z \leq 40$. These abundances can be used to address the questions surrounding the nature of the GCR source material and acceleration mechanism. Supernovae have long been thought to be responsible for accelerating the GCRs as they provide the power needed with a reasonable acceleration efficiency. There is evidence supporting the picture that the GCRs originate in superbubbles surrounding OB associations,[1,2] in which the source material arises from the outflows of Wolf-Rayet stars and from the ejecta of supernovae (SNII, SNIb,c) mixed with old interstellar gas. Two models have been proposed to explain the detailed composition of the GCR source material that is accelerated by supernova shocks. The first is based on the observation that GCR abundances are strongly enhanced over SS abundances based on first ionization potential (FIP),[3] which suggests that the GCR source might be an environment such as stellar atmospheres. The alternate model, based on volatility,[4,5] notes that most elements with low FIP also are refractory, i.e. have low condensation temperatures, which suggests that the GCR source could be enriched in material from interstellar dust grains.

TIGER data will improve the statistical precision in the measurement of the abundances of the elements $_{29}$Cu, $_{30}$Zn, $_{31}$Ga, $_{32}$Ge and $_{37}$Rb, which break the FIP-volatility degeneracy. Previous measurements of UH (Ultra-Heavy $Z \geq 30$) GCRs were made by the HEAO-3[6] and Ariel-6[7] satellite instruments, which were able to resolve the even-Z elements in the $30 \leq Z \leq 60$ range but did not resolve the odd-Z elements. The HEAO-C2 experiment[8] provided a measurement for the lower part of this range but with limited statistics. The ACE-CRIS experiment provided a measurement of the isotopic abundances for elements in the $29 \leq Z \leq 34$ range,[9] but with comparatively low statistics for the elemental abundances for $Z > 30$. Results from the 2001 TIGER flight[10] showed that the instrument has single charge resolution in the $30 \leq Z \leq 40$ range and yielded improved measurements of $_{30}$Zn, $_{31}$Ga and $_{32}$Ge, and the preliminary results of the combined 2001 and 2003 TIGER datasets[11] showed similar resolution and increased statistics in the UH range.

2. The TIGER instrument

TIGER is a Long Duration Balloon (LDB) borne experiment designed to operate in near vacuum that is capable of characterizing the charge and energy of the GCR with charges between $Z = 14$ (Silicon) and $Z = 40$ (Zirconium). The instrument, shown in Fig. 1, consists of four PVT scintillator radiators (St Gobains BC-416) read out with wavelength-shifter-bars (WLSB) (St Gobains BC-482A), two Cherenkov radiators in light collection boxes (one acrylic and one aerogel), and a scintillating optical fiber hodoscope. The radiators are arranged with two scintillators (S1 & S2) on top with a hodoscope plane (HT) in between. The two Cherenkov detectors are in the middle with the aerogel (C0) being above the acrylic (C1), and finally the other two scintillators (S3 & S4) with a hodoscope plane (HB) in between at the bottom. The light produced in the radiators and the hodoscope is measured using photomultiplier tubes (PMTs). The scintillators each have eight Hammamatsu R1924 PMTs mounted at the ends of the WLSBs that surround the radiator edges. The light collection boxes of the two Cherenkov radiators each have six Burle S83006F PMTs along each of their four edges. A total of 112 Hammamatsu R1924 PMTs are used in the two hodoscope planes. The signals from the PMTs are pulse height analyzed by the flight electronics.

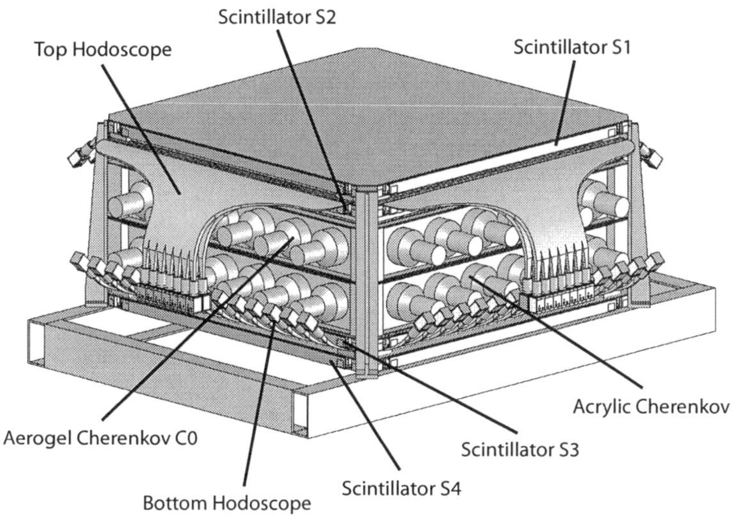

Fig. 1. The TIGER instrument.

The scintillators provide a measurement of light emitted as a function of path length traversed by the ionizing particle, dL/dx. The light produced is not directly proportional to the energy deposited due to saturation effects in the scintillator, which must be corrected for to determine the energy loss as a function of path length, dE/dx. The scintillators are also used in flight for the event trigger by requiring coincidence in top and bottom scintillators to ensure the particle is in the detector's geometry, as well as to determine which events met a minimum signal threshold for recording and transmission to ground. In post flight analysis, the top and bottom scintillators are also used to eliminate events that may have interacted within the instrument.

The Cherenkov radiators measure the velocity of the incident particles and contribute to their charge measurement. Cherenkov radiation is produced by a particle traversing a medium with a velocity greater than the speed of light within that medium, and it is proportional to the square of the particle's charge (Z) and is a function of the particle's velocity. Two different Cherenkov radiators are used since their different indices of refraction, 1.5 for the acrylic and 1.04 for the aerogel, provide different energy thresholds, 0.32 and 2.5 GeV/nucleon respectively. Together, these radiators provide TIGER with energy sensitivity between 0.3 and 10 GeV/nucleon in the instrument.

The scintillating fiber hodoscope measures the trajectory of particles through the TIGER instrument. The hodoscope has two planes, each consisting of two layers of perpendicular fibers formatted into tabs of six 1 mm square fibers. The tabs are formatted to 14 PMTs at either end, with a fine side receiving every fourteenth tab and a coarse side receiving groups of 14 consecutive tabs. This coding[12] allows for the determination of particle coordinates to within 6 mm for particles lighting only one tab and to within 3 mm for particles lighting more than one tab. The coordinates determined in the hodoscope layers are used to determine the angle of incidence of the particles and where they traverse the radiators. This allows corrections to be made for pathlengths and area effects within the radiators.

3. Results of two flights

TIGER has had two successful flights from McMurdo, Antarctica for a total of nearly 50 days of flight time. The December 21, 2001 – January 21, 2002 flight, lasting 31.8 days, had an average altitude of 118,800 ft (36,200 m) and 5.5 mbar of residual atmosphere. The altitude varied considerably over the duration of this flight from a high near 129,000 ft (39,300 m) at the beginning to a low near 109,000 ft (33,200 m) at the end due to a slow leak in the balloon. The December 17, 2003 – January 4, 2004 flight, lasting 18 days, had an average altitude 127,800 ft (39,000 m) and 4.1 mbar of residual atmosphere, with the altitude varying between a minimum of 121,000 ft (36,9000 m) and a maximum of 134,000 ft (40,800 m). There are $\sim 2/3$ as many high Z events from the 2003 flight as the 2001 due to the higher average altitude and a slight reduction in the amount of material in beam even though the flight was $\sim 1/2$ as long in duration.

Fig. 2 shows crossplots for a sample of events from the 2003 dataset. The plot on the left shows the sum of the top scintillator signals (S1 + S2) plotted versus the sum of the acrylic Cherenkov signal (C1), which are used to determine Z for particle energies below the threshold of the aerogel Cherenkov (C0). We see that there is good separation between the charge contours for each element with the exception of the high energy nuclei to the

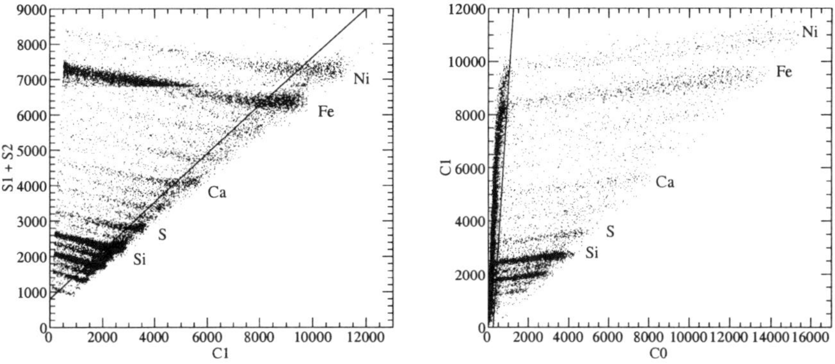

Fig. 2. Crossplots of top scintillator signals versus acrylic Cherenkov signal (left) and acrylic Cherenkov signal vs aerogel Cherenkov signal (right) for the 2003 dataset with every 100th event below S1 + S2 = 6800 and every 10th above plotted to show Ni.

right of the line. There is a small relativistic rise in the scintillation signal resulting in charge identification ambiguity in this energy region. The plot on the right shows the acrylic Cherenkov signal plotted versus the aerogel Cherenkov signal, which is used to assign charge (Z) to particles above the C0 threshold. The particles to the right of the line, which show good charge resolution, are most of the particles to the right of the line in the left panel. Thus we have used C0 to resolve the ambiguity in charge assignment of the higher energy particles ($E > 2.5$ GeV/nucleon) which results from a measurement of S and C1 only.

4. Preliminary analysis

The results of the preliminary analysis of the combined data from the 2001 and 2003 flights is shown in Fig. 3. On the left is a charge histogram of the combined data set with a 1000 times change in scale at $Z = 29$. We see that the charge resolution is good and that clear peaks are observed for $Z = 30$, 31, 32, and 34, along with the beginnings of low statistics peaks for $Z = 36$ and 38. The plot on the right compares the measured abundances relative to Ni/1000 with those of solar system source abundances modified by either FIP or Volatility and propagated to balloon altitude. The measured data have 1–σ statistical error bars, and are seen to generally agree with the model predictions within these errors. Where the statistical precision of the measured data is sufficient to discern between the two models the results are contradictory. The measured relative abundance of $_{31}$Ga agrees with the

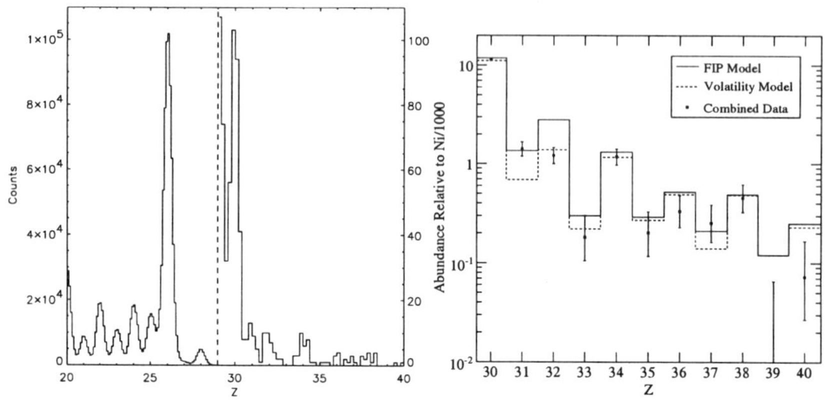

Fig. 3. Charge histogram of combined dataset from 2001 and 2003 flights (left) and comparison of the measured relative ultra-heavy abundances with 1–σ statistical error bars with FIP and Volatility model abundances propagated to balloon altitude (right).

FIP model, while that of $_{32}$Ge agrees with the Volatility model. It is possible that the semi-empirical cross sections used in the propagation models are incorrect and are responsible for these results, but if this is not the case this suggests that the source medium does not have simple solar system abundances, which may be expected for a superbubble source environment.

Acknowledgments

This research was supported by the National Aeronautics and Space Administration under grant NNG05WC04G.

References

1. J.C. Higdon and R.E. Lingenfelter, Ap.J., 590 (2003) 822.
2. W.R. Binns *et al.*, Ap.J., 634 (2005) 351.
3. M. Casse and P. Goret, Ap.J., 221 (1978) 860.
4. R.I. Epstein, MNRAS, 193 (1980) 723.
5. J.-P. Meyer *et al.*, Ap.J., 487 (1997) 182.
6. W.R. Binns *et al.*, Ap.J., 346 (1989) 997.
7. P.H. Fowler *et al.*, Ap.J., 314 (1987) 739.
8. J.J. Engelmann *et al.*, A&A, 233 (1990) 96.
9. J.S. George *et al.*, 26th ICRC, Salt Lake City (1999) OG 1, 13.
10. J. Link *et al.*, 28th ICRC, Japan (2003) OG 1, 1781.
11. S. Geier *et al.*, 29th ICRC, India (2005) OG 1, 93.
12. D.J. Lawrence *et al.*, NIM-A, 420 (1999) 402.

ISOTOPIC MASS SEPARATION WITH THE RICH DETECTOR OF THE AMS EXPERIMENT

LUÍSA ARRUDA, F.BARÃO, J.BORGES, F.CARMO,
P.GONÇALVES, R.PEREIRA M.PIMENTA
LIP/IST
Av. Elias Garcia, 14, 1º andar
1000-149 Lisboa, Portugal
e-mail: luisa@lip.pt

A. KEATING
ESTEC/ESA, Netherlands

The Alpha Magnetic Spectrometer (AMS) to be installed on the International Space Station (ISS) will be equipped with a proximity focusing Ring Imaging Čerenkov detector (RICH). Reconstruction of the Čerenkov angle and the electric charge with RICH are discussed. A likelihood method for the Čerenkov angle reconstruction was applied leading to a velocity determination for protons with a resolution around 0.1%. The electric charge reconstruction is based on the counting of the number of photoelectrons and on an overall efficiency estimation on an event-by-event basis. The isotopic mass separation of helium and beryllium is presented.

1. The AMS02 and the RICH detector

AMS (Alpha Magnetic Spectrometer) [1, 2] is a precision spectrometer designed to search for cosmic antimatter, dark matter and to study the relative abundances of elements and isotopic composition of the primary cosmic rays. It will be installed in the International Space Station (ISS), in 2008, where it will operate for a period of at least three years. It will be equipped with a Ring Imaging Čerenkov detector (RICH). This detector was designed to measure the velocity of singly charged particles with a resolution $\Delta\beta/\beta$ of 0.1%, to extend the electric charge separation up to the iron element, to contribute to the albedo rejection and to contribute to the e/p separation.

The RICH of AMS is a proximity focusing Čerenkov radiation detector. Its radiator is composed by aerogel (n=1.05) and a sodium fluoride (NaF

1.334) squared region placed at the center and covering an acceptance of ~10%. The whole detector set will be covered by a high reflectivity conical mirror increasing the reconstruction efficiency. Photons will be detected in a matrix with 680 photomultipliers (PMTs) coupled to light guides. There will be a large non-active area at the center of the detection area due to the insertion of an electromagnetic calorimeter. For a more detailed description of the RICH detector see reference [3]. Figure 1 shows a view of the RICH and a beryllium event display with a view of the PMT detailed matrix.

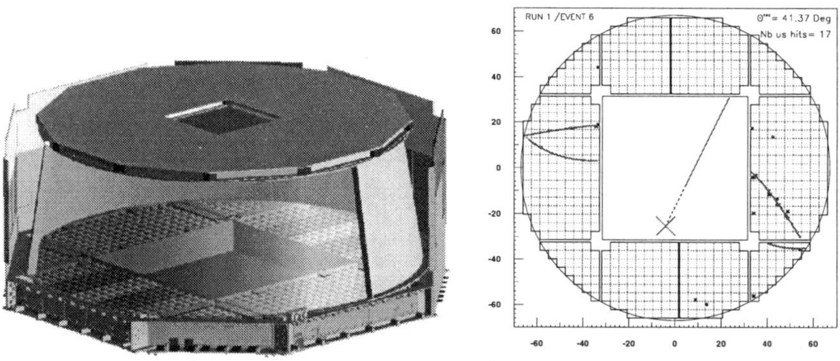

Figure 1. *On the left:* View of the RICH detector. *On the right:* Beryllium event display generated in a NaF radiator. The reconstructed photon pattern (full line) includes both reflected and non-reflected branches. The outer circular line corresponds to the lower boundary of the conical mirror. The square is the limit of the non-active region.

2. Velocity reconstruction

A charged particle crossing a dielectric material of refractive index n, with a velocity β, greater than the speed of light in that medium emits photons. The aperture angle of the emitted photons with respect to the radiating particle track is known as the Čerenkov angle, θ_c, and it is given by (see [4]):

$$\cos\theta_c = \frac{1}{\beta\, n} \tag{1}$$

It follows that the velocity of the particle, β, is straightforward derived from the Čerenkov angle reconstruction, which is based on a fit to the pattern of the detected photons. Complex photon patterns can occur at the detector plane due to mirror reflected photons, as can be seen on right display of Figure 1. The event displayed is generated by a simulated beryllium nuclei in a NaF radiator.

The Čerenkov angle reconstruction procedure relies on the information of the particle direction provided by the tracker. The tagging of the hits signaling the passage of the particle through the solid light guides in the detection plane, provides an additional track element, however, those hits are excluded from the reconstruction. The best value of θ_c will result from the maximization of a likelihood function, built as the product of the probabilities, p_i, that the detected hits belong to a given (hypothesis) Čerenkov photon pattern ring,

$$L(\theta_c) = \prod_{i=1}^{nhits} p_i^{n_i} \left[r_i(\theta_c) \right] . \tag{2}$$

Here r_i is the closest distance of the hit to the Čerenkov pattern. For a more complete description of the method see [5]. The resolution achieved for protons of 20 GeV/c/nuc is ~4 mrad. The evolution of the relative resolution of beta with the charge can be observed on the left plot of Figure 2. It was extracted from reconstructed events generated in a test beam at CERN in October 2003 with fragments of an indium beam with a momentum per nucleon of 158 GeV/c/nuc, in a prototype of the RICH detector.

3. Charge reconstruction

The Čerenkov photons produced in the radiator are uniformly emitted along the particle path inside the dielectric medium, L, and their number per unit of energy depends on the particle's charge, Z, and velocity, β, and on the refractive index, n, according to the expression:

$$\frac{dN_\gamma}{dE} \propto Z^2 L \left(1 - \frac{1}{\beta^2 n^2}\right) = Z^2 L \sin^2 \theta_c \tag{3}$$

So to reconstruct the charge the following procedure is required:

- Čerenkov angle reconstruction.
- Estimation of the particle path, L, which relies on the information of the particle direction provided by the tracker.
- Counting the number of photoelectrons.
  The number of photoelectrons related to the Čerenkov ring has to

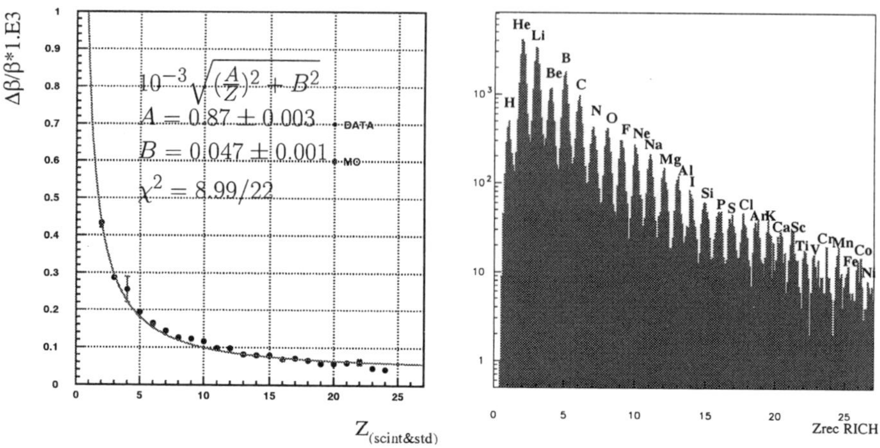

Figure 2. At left evolution of the relative resolution on beta with the charge and at right the reconstructed charge peaks. Both are reconstructions with data from a test beam at CERN in October 2003, using an indium beam of 158 GeV/c/nuc.

be counted within a fiducial area, in order to exclude the uncorrelated background noise. Therefore, photons which are scattered in the radiator are excluded. A distance of 15 mm to the ring was defined as the limit for photoelectron counting, corresponding to a ring width of ~4 pixels.

- Evaluation of the photon detection efficiency. The number of radiated photons (N_γ) which will be detected ($n_{p.e.}$) is reduced due to the interactions with the radiator (ε_{rad}), the photon ring acceptance (ε_{geo}), light guide (ε_{lg}) and photomultiplier efficiency (ε_{pmt}).

$$n_{p.e.} \sim N_\gamma \, \varepsilon_{rad} \, \varepsilon_{geo} \, \varepsilon_{lg} \, \varepsilon_{pmt} \tag{4}$$

The charge is then calculated according to expression 3, where the normalization constant can be evaluated from a calibrated beam of charged particles. In the right plot of Figure 2 are visible reconstructed charge peaks from the mentioned test beam at CERN in October 2003. These results were obtained with aerogel radiator 1.05 and 2.5 cm thick. A charge resolution for helium events slightly better than $\Delta Z \sim 0.2$ was observed together with a systematic of 1%. A clear charge separation up to Z=27 was achieved. For a more complete description of the charge reconstruction method see [5].

4. Isotopic element separation

Isotopic separation and particularly the ratios $^3He/^4He$ and $^{10}Be/^9Be$ is a major part of the physics goals where the RICH plays a fundamental role within AMS. The presence of a mixed radiator with a NaF radiator at the center will allow AMS to cover a kinematic energy range from 0.5 GeV/nucleon up to around 10 GeV/nucleon.

Samples of helium and beryllium nuclei corresponding to 1 day and 1 year of data taking, respectively, were simulated. These samples were generated according to [6] for helium and [7] for beryllium nuclei. Afterwards, the spectra was modulated taking into account the geomagnetic field. The masses were reconstructed using a momentum uncertainty $\sim$2%.

The reconstructed masses were fitted with a sum of two gaussian functions:

$$f(m) \propto \alpha \left(G_1(M_1, \sigma_1) + G_2(M_2, \sigma_2)\right)$$

where M_i, σ_i and α are respectively the isotopic mass central value, the mass width and the relative weight of the two distributions.

Figure 3 presents the isotopic ratios obtained from the fits as function of the kinetic energy. Isotopic ratios from events crossing the sodium fluoride radiator are fairly measured up to the aerogel threshold. From there on, the aerogel allows to measure the isotopic ratios up to around 10 GeV/nucleon of kinetic energy. Above 10 GeV/nuc the mass relative resolution is greater than 8.5% for He and greater than 6% for Be.

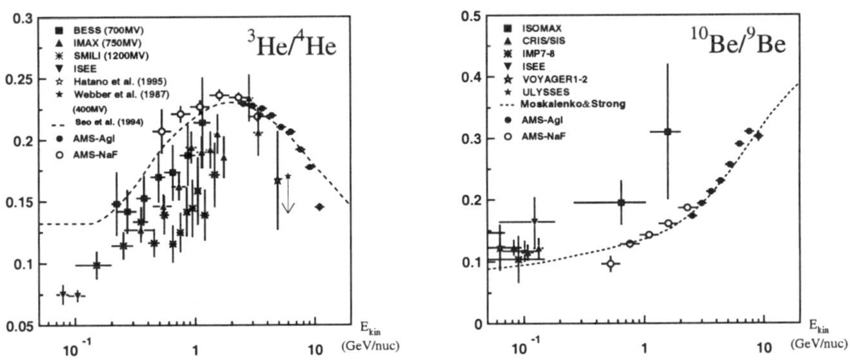

Figure 3. Reconstructed isotopic ratios of helium and beryllium simulated events as function of kinetic energy per nucleon. The aerogel in study has a refractive index of 1.050.

5. Conclusions

AMS is a spectrometer designed for antimatter, dark matter searches and for measuring relative abundances of nuclei and isotopes. The instrument will be equipped with a proximity focusing RICH detector based on a mixed radiator of aerogel and sodium fluoride, enabling velocity measurements with a resolution of about 0.1% and extending the charge measurements up to the iron element. Velocity reconstruction is made with a likelihood method. Charge reconstruction is made in an event-by-event basis. Both algorithms were successfully applied to simulated data samples with flight configuration. Evaluation of the algorithms on real data taken with the RICH prototype was performed at the LPSC, Grenoble in 2001 and in the test beam at CERN, in October 2002 and 2003. The RICH radiator will allow AMS to perform helium and beryllium isotopic separation up to 10 GeV/nucleon.

References

1. S. P. Ahlen, V. M. Balebanov et al, Nucl. Instrum. Methods A **350**, 34 (1995).
2. V. M. Balebanov, AMS proposal to DOE (1995).
3. M.Buenerd. Proceedings of the Fourth Workshop on Rich Detectors (RICH02) June 5-10, 2002, Pylos, Greece.
4. T.Ypsilantis and J.Seguinot, Nucl. Instrum. Methods A **343**, 30 (1994).
5. F.Barão, Nucl. Instrum. Methods A **502**, (2003).
6. E.S. Seo, ApJ **431**, 705 (1994).
7. A.W. Strong and I.V. Moskalenko, ApJ **509**, 212 (1998).

MULTIDIRECTIONAL MUON TELESCOPES AND eEAS ARRAYS FOR HIGH ENERGY COSMIC RAY RESEARCH

LEV I. DORMAN[1,2]

[1] *Israel Cosmic Ray and Space Weather Center and Emilio Sègre Observatory affiliated to Tel Aviv University, Technion and Israel Space Agency, Israel*
E-mail: lid@physics.technion.ac.il [2] *Cosmic Ray Department of IZMIRAN, Russian Academy of Science, Russia*

Two multidirectional muon telescopes with EAS arrays are now under construction in Israel: one from 24 scintillators on Mt. Hermon (in combination with neutron monitor), and one from 96 scintillators as semi-underground (in the big bomb-shelter in Qazrin at a distance of about 1 nkm from the Central Laboratory of the Israel Cosmic Ray & Space Weather Center). The big one consists from 49 scintillation detectors inside the special constructed building with very light roof over the bomb-shelter and 49 scintillation detectors underground inside the bomb-shelter. This multidirectional telescope contain more than two thousand elementary telescopes directed at different zenith and azimuthal angles and formed by double coincidences of any top scintillator with each bottom scintillator (the effective energy of primary CR from about 50 GeV for vertical direction to about 1-2 TeV for very inclined directions). It will give possibility to investigate global and other types of galactic CR modulations in the Heliosphere at very high energies, near the upper limit of CR energy on which magnetic fields frozen in solar wind may yet influence. Also we plane to obtain detailed information on the sidereal CR anisotropy in this range of energy.

We will measure also three types of EAS. Our estimations show that by EAS array we can continue measure high energy CR time variations in the broad range from about 1-2 TeV to about 10,000 TeV. By this experiment, we suppose to investigate with a high accuracy CR anisotropy in the Galaxy in dependence of particle energy and CR modulation in the Heliosphere at high-energy range.

1. Introduction

The Israel Cosmic Ray and Space Weather Center (ICR&SWC) and Israeli - Italian Emilio Segre' Observatory (ESO) were established in 1998, with affiliation to Tel Aviv University, to the Technion (Israel Institute of Technology, Haifa) and to the Israel Space Agency. The mobile CR Neutron Monitor was prepared by the collaboration of Israeli scientists of ICR&SWC

/ESO with Italian scientists of CR Group of Roma-Tre University and of the Cosmic Radiation Sector IFSI/CNR and transferred in June 1998 on Mt. Hermon (33°18' N, 35°47.2' E, 2055 m above sea level, vertical cut off regedity $R_c = 10.8$ GV. The results of measurements (data taken at one minute intervals of CR neutron total intensities at two separate 3NM-64 sections, as well as similar one minute data about the intensities relating to neutron multiplicities m = 1, 2, 3, 4, 5, 6, 7 and $\geq$8) are stored in the computer. Similar one minute data relating to the atmospheric electric field, wind speed, three components of geomagnetic field, air temperature outside, and humidity and temperature inside the CR

Observatory are also recorded and archived. Each month one hour data of ESO are sent to the World Data Center in Boulder (USA, Colorado) and to many CR Observatories in the world. An automatic electric power supply using Uninterruptible Power Supply (UPS) and a diesel generator guarantees continuous power for ESO. There is a direct radio connection in real time from ESO on Mt. Hermon to the Central Laboratory of ICR&SWC in Qazrin, and to the Internet. To extend the experimental basis of ICR&SWC/ESO on Mt. Hermon (see in Dorman [1]) and a great semi-underground plastic scintillation multidirectional muon telescope are now under construction, with more than two thousand two-coincidences channels for vertical and inclined directions at different zenith and azimuthal angles together with EAS installation in a former bomb shelter in Qazrin, which will be described shortly below.

1.1. *Description of semi-underground multi-directional muon telescope*

A semi-underground multi-directional muon telescope is presently under construction. Figure 1 depicts the planned underground multidirectional muon telescope that will start to work in Qazrin in near future.

2. Description of EAS array combined with semi-underground multi-directional muon telescope in Qazrin

We will measure three types of EAS. The first type is formed by coincidences in different combinations of 2-fold, 3-fold, 4-fold, and so on (up to 49-fold) only between top 49 scintillation detectors. The second type – the same but using coincidences only between bottom 49 scintillation detectors inside underground bomb-shelter. By the comparison and combinations of the

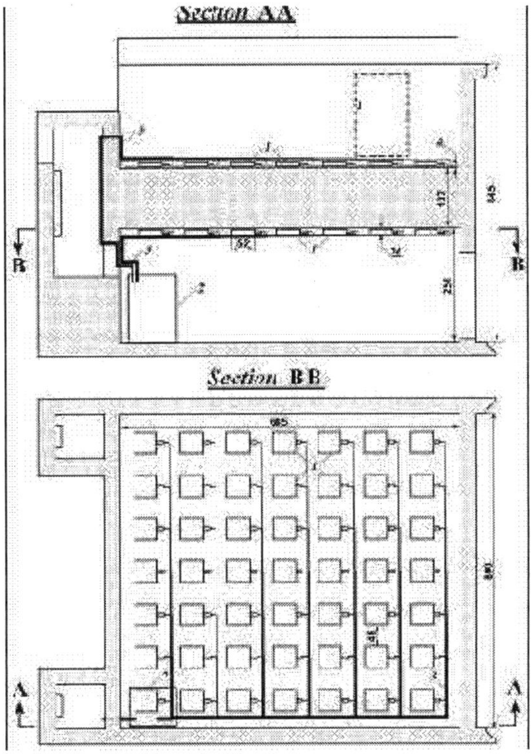

Fig. 1. The disposition of the semi-underground multidirectional muon telescope in the in a former bomb-shelter in Qazrin, Israel (with characteristics partly listed in Table 1). 1- scintillation detectors (see Figure 1); 2- acquisition system and computers; 3 – connection and power feeders. Dimensions are given in cm.

first and second types of EAS we will select EAS in dependence of the ratio muons/electrons to separate showers generated by high energy gamma-rays, protons or heavy particles. The third type will be formed by coincidences (with some time lag) in different combinations of 2-fold, 3-fold, 4-fold, and so on of elementary muon telescopes in the same direction for detection inclined muon showers. Our estimations show that by EAS array we can continue measure high energy CR time variations in the broad range from about 1-2 TeV to about 10,000 TeV.

By coincidences in different combinations of upper scintillators, we obtain counting of EAS (electron-photon component). This array can be considered as local because the distances between detectors are much smaller than the effective radius of EAS on the level of observations. In this case,

No.	Channel	$\phi,\ ^o$	$\theta,\ ^o$	#	Rate
1	Vertical		0	49	1782878
2	N1, W1, S1, E1	0, 90, 180, 270		42*4	574890*4
3	NW1, SW1, SE1, NE1	45, 135, 225, 315	41.3	36*4	236011*4
4	N2, W2, S2, E2	0, 90, 180, 270	51.1	35*4	77619*4
5	NW2, SW2, SE2, NE2	45, 135, 225, 315	60.3	25*4	13373*4
6	N3, W3, S3, E3	0, 90, 180, 270	61.8	28*4	11427*4
7	N4, W4, S4, E4	0, 90, 180, 270	68.1	21*4	2078*4
8	NW3, SW3, SE3, NE3	45, 135, 225, 315	69.2	16*4	1165*4
9	N5, W5, S5, E5	0, 90, 180, 270	72.1	14*4	424*4
10	N4W4, S4W4, S4E4, N1E	45, 135, 225, 315	74.1	9*4	138*4
11	N6, W6, S6, E6	0, 90, 180, 270	75.0	7*4	78*4
12	N5W5, S5W5, S5E5, N5E5	45, 135, 225, 315	77.2	4*4	18*4
	Total in 45 directions			997	5451767

on passage of EAS of density ρ (mean number of particles per 1 m^2) the probability that not a single particle will pass through a detector of effective area σ will be $\exp(-\rho\sigma)$. The probability of at least one particle crossing through a detector will be $\omega = 1 - \exp(-\rho\sigma)$. The particle distribution in EAS may be represented in the form $\rho(r) = u(r)\, N_e$, where r is the distance from the EAS axes, N_e is the total number of particles in electron-photon component of EAS, and $u(r)$ is the function satisfying the normalization condition and has the form:

$$2\pi \int_0^\infty u(r)\, r dr = 1$$

with

$$u(r) = \begin{cases} ar^{-1} \exp(-r/r_o) & \text{if} r \leq r_o, \\ u(r) = br^{-2.6} & \text{if} r \geq r_o. \end{cases} \tag{1}$$

Here r_o is the effective radius of the shower (r_o= 55 m for sea level observations, and r_o= 80 m for mountain observations on the level about 3 km). Coefficients a and b are determined from the condition of normalizing and of tie-in of the function $u(r)$ at the point r = r_o. It gives on the basis of Eq. (1):

$$\begin{aligned} a &= er_o^{-1}\left[2\pi\left(e\left(1 - 1/e\right) + 1/0.6\right)\right]^{-1} = 0.12781 \times r_o^{-1} \\ b &= r_o^{0.6}\left[2\pi\left(e\left(1 - 1/e\right) + 1/0.6\right)\right]^{-1} = 0.04702 \times r_o^{0.6} \end{aligned} \tag{2}$$

Eq. (2) gives for the altitude 3 km (r_o= 80 m) a = 1.598×10^{-3} m^{-1}, b = 0.6518 $m^{0.6}$; for Mt Hermon (altitude 2 km, r_o= 72 m) a = 1.775×10^{-3} m^{-1}, b = 0.612 $m^{0.6}$; for sea level (r_o= 55 m) a = 2.332×10^{-3} m^{-1}, b = 0.5206 $m^{0.6}$.

Let us suppose that the axis of EAS with total number of particle N_e crossed the observation level in some point P and actuated simultaneously any n detectors of array with total m detectors (meaning that through each of these n detectors crossed at least one particle from total number $N_e$in EAS) and not actuated other any m – n detectors. Let the distance from point P to the detector i is r_i (i = 1, 2, ... m). In this case the probability to detect this EAS will be

$$\omega_{nm}(P) = C_m^n \prod_{i=1}^{n} (1 - \exp(-\rho_i \sigma)) \prod_{i=n+1}^{m} \exp(-\rho_i \sigma)$$
$$\approx C_m^n (1 - \exp(-u(r) N_e \sigma))^n \exp(-(m-n) u(r) N_e \sigma) , \quad (3)$$

where $C_m^n = m!(n!(m-n)!)^{-1}$. In Eq. (3) we take into account that in our case the distances between detectors $<< r_o$, therefore we can put for all $r_i \approx r$, where r is the distance from point P to the center of installation. Let us take into account also that $N_e \approx 0.3E_o^s$, where E_o is the energy of primary particle in GeV, and $s \approx 1.1$ at mountain level 3 km, and $s \approx 1.2$ at sea level. Because P may be elsewhere, the total probability to detect EAS with N_e particles (generated by primary particle with energy E_o) simultaneously by any n detectors of the local installation with total m detectors will be in the unity of time

$$\omega_{nm}(N_e, r_o, \sigma) dN_e \approx 2\pi C_m^n D(N_e) dN_e \times$$
$$\times \int_0^\infty (1 - \exp(-u(r) N_e \sigma))^n \exp(-(m-n) u(r) N_e \sigma) r dr; \quad (4)$$

$$\omega_{nm}(E_o, r_o, \sigma) dE_o \approx 2\pi C_m^n D(E_o) dE_o \times$$
$$\times \{ \int_0^{r_o} (1 - \exp(-0.03834 r_o^{-1} r^{-1} \exp(-r/r_o) E_o^s \sigma))^n \times$$
$$\times \exp\left(-0.03834 r_o^{-1} r^{-1} (m-n) \exp(-r/r_o) E_o^s \sigma\right) r dr +$$
$$+ \int_{r_o}^\infty (1 - \exp(-0.01411 r_o^{0.6} r^{-2.6} E_o^s \sigma))^n \times$$
$$\exp\left(-0.01411 r_o^{0.6} r^{-2.6} (m-n) E_o^s \sigma\right) r dr \} \quad (5)$$

where $D(N_e)$ and $D(E_o)$ are differential spectrums of EAS, and E_o is in GeV, r_o is in m, σ is in m^2. The expected counting rate $I_{nm}(r_o,\sigma)$(i.e. number of detected EAS per unity of time by any n coincidences from total m detectors with effective area σ each) will be

$$I_{nm}(r_o,\sigma) = \int_0^\infty \omega_{nm}(N_e,r_o,\sigma)\,dN_e = \int_0^\infty \omega_{nm}(E_o,r_o,\sigma)\,dE_o\,. \tag{6}$$

The coupling functions characterized the sensitivity of installation to detect EAS will be

$$W_{nm}(N_e,r_o,\sigma) = \omega_{nm}(N_e,r_o,\sigma)\,/I_{nm}(r_o,\sigma)\,;\; W_{nm}(E_o,r_o,\sigma) = \omega_{nm}(E_o,r_o,\sigma)\,/I_{nm}(r_o,\sigma)\,. \tag{7}$$

It is easy to see that these coupling functions are normalized for any E_o, σ, n and m:

$$\int_0^\infty W_{nm}(N_e,r_o,\sigma)\,dN_e = \int_0^\infty W_{nm}(E_o,r_o,\sigma)\,dE_o = 1\,. \tag{8}$$

The differential energy spectrum of primary CR can be represented by (in m^{-2}sec^{-1}GeV^{-1}sr^{-1}):

$$D(E_o) = \begin{cases} 2.2\times10^4\times E_o^{-2.7}, & if\;\; E_o \le 3\times10^6 GeV \\ 3.8105\times10^7\times E_o^{-3.2}, & if\;\; E_o \ge 3\times10^6 GeV \end{cases} \tag{9}$$

The expected dependence of counting rate in dependence of n is shown in Figure 2.

3. Multi-directional muon telescope and EAS installation on Mt. Hermon

On Mt. Hermon in the Emilio Segre' Observatory of the Israel Cosmic Ray and Space Weather Center, a plastic scintillation multidirectional muon telescope is under construction, with 144 two-coincidence channels in combination with NM-IQSY. We calculate the angle diagrams of telescope sensitivity and coupling functions for many directions at different zenith and azimuthal angles. We give a description of the electronics scheme and registration system. We also discuss possibly using the telescope in combination with the detection of the total neutron component and different neutron multiplicities. The scintillators of the telescope will be used also for continued measurement of electron-photon and muon EAS arriving in vertical and at different inclined directions. We also calculate the expected coupling

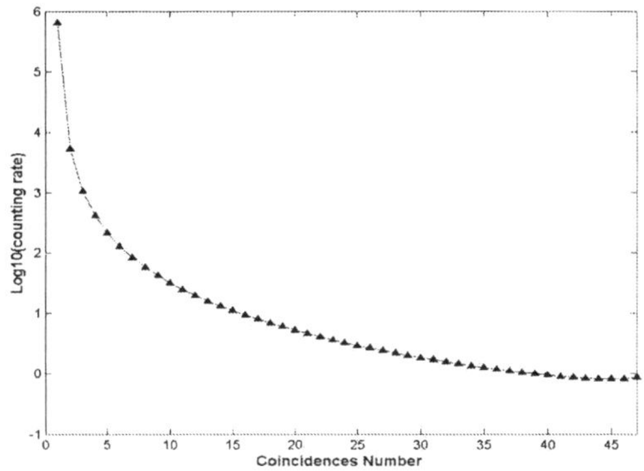

Fig. 2. The expected dependence of counting rate of EAS per sec in dependence of coincidences number n for the EAS installation in the bomb-shelter.

functions (or response) functions for different types of EAS. We consider also how NM-IQSY will work in combination with EAS installation for measuring of nucleonic component of EAS.

In Figure 3 we show a block scheme of the main components of the Emilio Segre' Observatory (ESO) on Mt. Hermon and their connection with the Central Laboratory of Israel Cosmic Ray and Space Weather Center in Qazrin and with the Internet.

Figure 4 depicts the cross-section of Emilio Segre' Observatory and disposition of both sections of NM-IQSY, as well as the multi-directional muon telescope to be installed in combination with NM-IQSY.

From Figure 4 it can be seen that multidirectional muon telescope consisted 24 scintillation detectors 50 cm $\times$ 50 cm $\times$ 10 cm (12 under NM and 12 above NM) with 144 double coincidences formed 12 vertical and 132 inclined telescopes at different zenith and azimuthally angles (see Table 2). By coincidences in different combinations of upper scintillators, we obtain a counting of EAS (electron-photon component). This array can be considered as local because the distances between detectors are much smaller than the effective radius of EAS on the level of observations. In this case, on the passage of EAS of density ρ (mean number of particles per 1 m^2) the probability that not a single particle will pass through detector of effective area σ will be

$$\exp(-\rho\sigma) \ .$$

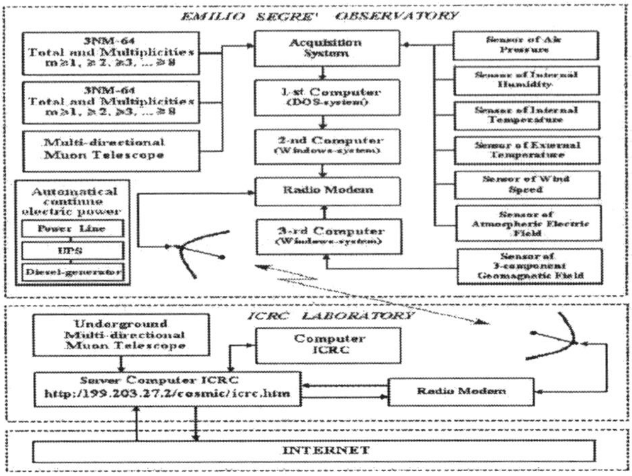

Fig. 3. Schematic description of the main components of the Israeli–Italian Emilio Segre' Observatory (ESO) and their connection with the Central Laboratory in Qazrin and with Internet. Also shown multi-directional muon telescope which will work in combination with NM-IQSY and EAS (see Figure 4).

The probability of at least one particle crossing through detector will be

$$\omega = 1 - \exp\left(-\rho\sigma\right) \ .$$

The particle distribution in EAS may be represented in the form

$$\rho\left(r\right) = u\left(r\right) N_e \ ,$$

where r is the distance from the EAS axes, N_e is the total number of particles in electron-photon component of EAS, and $u\left(r\right)$ is the function satisfying the normalization condition and has the form described by Eq. 1.

Then we can use Eq. 2-9, which we obtained for semi-underground installation, only now we take into account that the installation is on Mt. Hermon, altitude 2 km. In this case

r_o= 72 m, a = 1.775×10^{-3} m^{-1}, b = 0.6119 m$^{0.6}$ (10)

and m = 12.

The dependence of expected counting rate $I_{nm}\left(r_o, \sigma\right)$for Mt. Hermon (m = 12, σ = 0.25 m^2) from number of coincidences n is shown in Figure 5.

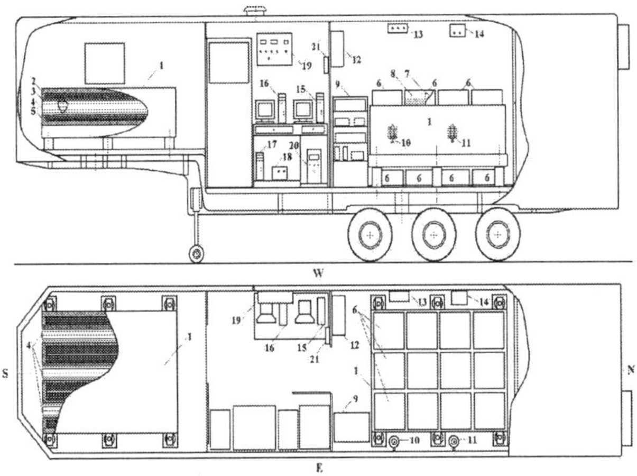

Fig. 4. Schematic view of the Israeli-?Italian Emilio Segre' Observatory (ESO) on Mt. Hermon, showing both sections of NM-IQSY and the multi-directional muon telescope (the lead of NM will be used as a shield for the muon telescope): the top – vertical cross-section; the bottom – view from above. 1 – two 3-counters sections of NM-IQSY; 2 – polyethylene plates; 3 – polyethylene tubes; 4 – neutron counter $^{10}BF_3$; 5 – lead tubes; 6 – scintillation detector; 7 – photo-multiplayer; 8 – plastic scintillator 50 cm×50 cm×10 cm; 9 –acquisition system; 10 – sensor of humidity inside Observatory; 11 - sensor of air temperature inside Observatory; 12 – the system of continuous electric power supply; 13 – radio-modem; 14 – recording instrument of EFS-1000 (for measurements of atmospheric electric field); 15 – 1-st computer (DOS-system); 16 – 2-nd computer (Windows system); 17 – 3-rd computer for geomagnetic field (Windows system); 18 - recording instrument for registration of the 3 components of geomagnetic field; 19 – control panel of electric power; 20 – UPS.

4. Acknowledgements

My great gratitude to Yu. Ne'eman, A. Sternlieb, Y. Israel, Z. Kaplan, L. Pustil'nik, I. Zukerman, and S. Applbaum, - for constant support, collaboration, and interesting discussions.

References

1. L.I. Dorman, *Cosmic Rays in the Earth's Atmosphere and Underground*, Kluwer, Dordrecht/Boston/London (2004).

No.	Channel	$\phi,^o$	$\theta,^o$	#	Rate
1	Vertical		0	12	1620000
2	N1, S1	0, 180	26.6	48*2	552960
3	W1, E1	90, 270	26.6	9*2	622080
4	NW1, SW1, SE1, NE1	45, 135, 225, 315	35.3	6*4	240000
5	N2, S2	0, 180	45.0	4*2	19999
6	W2, E2	90, 270	45.0	6*2	30000
7	N1W2, S1W2, S1E2, N1E2	63.4, 116.6, 243.4, 296.6	48.2	4*4	47407
8	N2W1, S2W1, S2E1, N2E1	26.6, 153.4, 206.6, 333.4	48.2	3*4	35555
9	NW2, SW2, SE2, NE2	45, 135, 225, 315	54.7	2*4	33750
10	W3, E3	90, 270	56.3	3*2	11797
11	N1W3, S1W3, S1E3, N1E3	71.6, 108.4, 251.6, 288.4	57.7	2*4	6298
12	N2W3, S2W3, S2E3, N2E3	56.3, 123.7, 236.3, 303.7	60.7	1*4	1759
13	Total in 45 directions			144	4094748

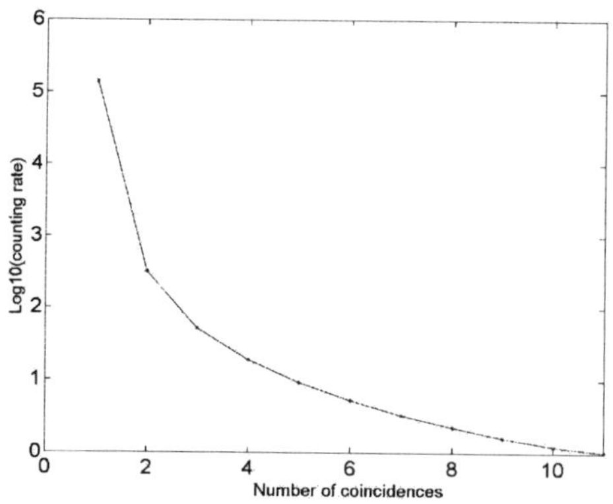

Fig. 5. Expected of EAS counting rate per sec for Mt Hermon installation.

gamma ray & neutrino astronomy

STUDY OF GALACTIC GAMMA RAY SOURCES WITH MILAGRO

Jordan A. Goodman for the Milagro Collaboration

Department of Physics, University of Maryland,
College Park, Maryland, 20742, USA
E-mail: goodman@umdgrb.umd.edu
www.physics.umd.edu

The diffuse gamma radiation arising from the interaction of cosmic ray particles with matter and radiation in the Galaxy is one of the few probes available to study the origin of the cosmic rays. Milagro is a water Cherenkov detector that continuously views the entire overhead sky. The large field-of-view combined with the long observation time makes Milagro the most sensitive instrument available for the study of large, low surface brightness sources such as the diffuse gamma radiation arising from interactions of cosmic radiation with interstellar matter. In this paper we report our results on diffuse emission from the galactic plane and in particular the Cygnus region. Our observations show at least one new TeV source MGRO J2020+37 as well as correlations with the matter density in the region as would be expected from cosmic-ray proton interactions. However, the TeV gamma-ray flux from the Cygnus region (after excluding MGRO J2020+37) is roughly 5 times that expected from a conventional model of cosmic ray production and propagation.

Keywords: Milagro; Galactic Sources; TeV Gamma-rays.

1. Introduction

Milagro is the first large area, continuously operating, water Cherenkov detector used for gamma-ray astronomy. Milagro consists of a 24 million liter water reservoir instrumented with 723 photomultiplier tubes (PMTs) surrounded by an array of 175 water tanks. The top layer of 450 PMTs is beneath 1.3 meters of water and is used to trigger the detector and to reconstruct the direction of the primary gamma ray (or cosmic ray). The bottom layer of 273 PMTs is beneath 6 meters of water and is used to measure the penetrating component of air showers induced by hadronic cosmic rays. The PMT spacing in the reservoir is 2.7 meters and the area enclosed is 4800 m^2. The surrounding array of water tanks is dispersed over 34,000 m^2. Each tank is cylindrical with a 1.6 meter radius and a depth of 1 meter. The tanks are instrumented with a single PMT located at the top of the tank looking down into the water volume. Milagro began physics operation in 2000 taking data with the central water reservoir. In 2004, construction of the outrigger array was completed. Before the installation of the outriggers the small size of the water reservoir limited the sensitivity of Milagro such that the Crab Nebula was observed at $\sim 4\sigma$ in one year of operation. The completion of the outriggers enabled a large increase in the sensitivity of the instrument, enabling us to detect the Crab Nebula at over 8 standard deviations in a single year of observation. This factor of ~ 2 increase in sensitivity (as shown in figure 1) has dramatically changed our view of the high-energy sky. It also means that the data currently being taken now with Milagro is substantially more important than our original data and we are not simply increasing our sensitivity by the square root of time over a 6-year observational period.

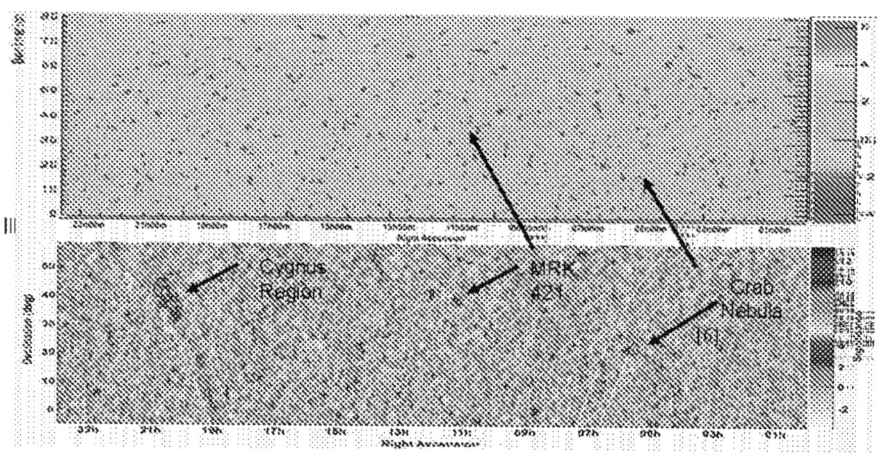

Fig. 1. The northern sky seen in TeV gamma rays using Milagro data. The top panel shows the sky as observed before the outriggers and the bottom panel after the completion of the outrigger array.

To date, our most important observational results with Milagro have been: the first detection of TeV gamma rays from the Galactic plane [1], the mapping of the diffuse Galactic gamma-ray emission at TeV energies, including the detection of the Cygnus Region at high-significance (over 10s) [2] , the discovery of a new (slightly extended) source of TeV gamma rays embedded in the Cygnus Region [3], and the possible detection of a gamma-ray burst with our prototype instrument Milagrito [4]. In addition, we have detected TeV gamma rays from the active galaxies Mrk 501 [5], Mrk 421 [6] plus the Crab Nebula [7], set stringent upper limits on the prompt TeV emission from several gamma ray bursts [8], and performed the most sensitive survey of the northern hemisphere at TeV energies [6]

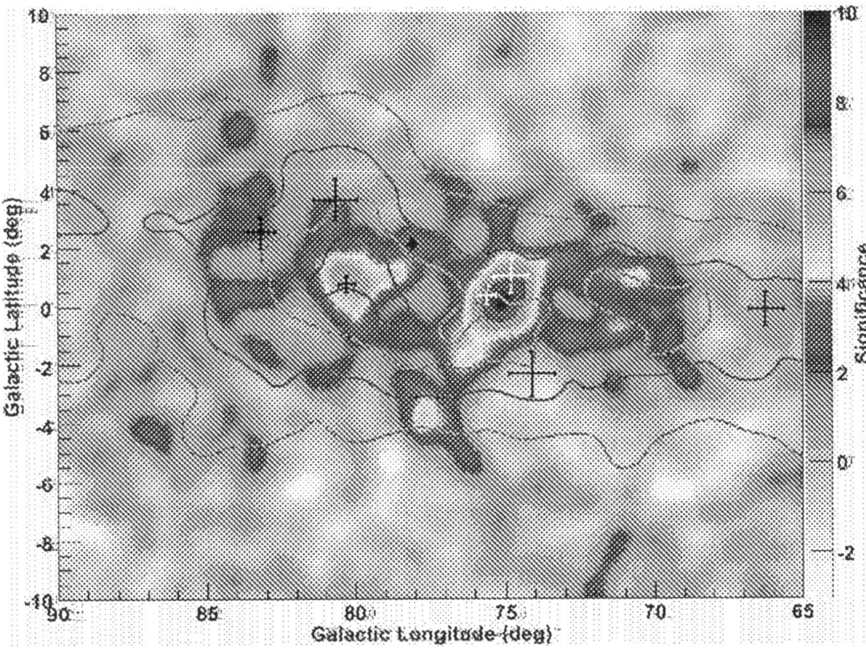

Fig. 2. The Cygnus Region of the Galaxy as seen in TeV gamma rays. Superimposed on the image are contours showing the matter density in the region. The probability that the two distributions are not correlated is 1.5x10-6. The crosses show the location of the EGRET sources and their corresponding location errors.

2. Survey of the Northern Hemisphere

The sensitivity of Milagro has been dramatically improved with the addition of the outrigger array. The effect of this gain in sensitivity is best demonstrated by comparing the results of our published sky survey [1] which used approximately 1000 days of data (top of figure 1) and our current data set of just over 2000 days of data (bottom of figure 1). While both the Crab Nebula and the active galaxy Mrk421 are visible in the top panel, the improvement since the outriggers is dramatic. The significance of the Crab

Nebula has increased from 6σ to 14σ and the Galactic plane is now clearly visible, even in this broad map. Our current sensitivity is 8σ per year for a 1 Crab flux. (It should be noted that for a source harder than the Crab, our sensitivity improves).

3. The Cygnus region and the discovery of a TeV gamma-ray source

In 2005, we published the first detection of diffuse TeV gamma-ray emission from the inner Galaxy [1]. The flux of the Milagro detection is not consistent with expectations from cosmic ray interactions if the local cosmic ray flux is indicative of the flux in the rest of the galaxy. There are several possible explanations for this excess: the local cosmic ray flux is unusually low, the local spectral index is soft relative to the rest of the Galaxy, and the existence of unresolved point sources. With the more recent data, taken since the completion of the outriggers, we have refined the analysis to investigate the Cygnus Region of the Galaxy in more detail. The Cygnus Region of the Galaxy is a natural laboratory for the study of cosmic ray origins. It contains a large column density of interstellar gas that should lead to strong emission of diffuse gamma rays and is also the home of potential cosmic-ray acceleration sites (Wolf-Rayet stars [9], OB associations [10], and supernova remnants [11]). Figure 2 shows a detailed view of the Cygnus Region in TeV gamma rays. Superimposed on the figure are contour lines indicating the matter density in the region and the location (and location errors) of the EGRET sources (all unidentified) in the region. There is definitive evidence for a new source of TeV gamma rays (MGRO J2019+37). MGRO J2019+37 is detected at over 10 standard deviations and its location is consistent with the location of two EGRET sources one of which has been tentatively identified with a pulsar wind nebula [12]. Though it is not evident from this figure, this source is most likely extended with a width of 0.32±0.12 degrees. The best-fit location of this source is R.A.=$304.83° \pm 0.14°$stat $\pm 0.3°$sys and Dec.= $36.83° \pm 0.08°$stat $\pm 0.25°$sys. Assuming a differential source spectrum of $E^{-2.6}$, the Milagro flux measurement at the median energy of 12 TeV is given by E^2 dN/dE = (3.49 ± 0.47stat $\pm$ 1.05sys) x 10^{12} TeV cm^{-2} s^{-1}. A change in the assumed source spectral index from -2.6 to -2.7 changes the integral flux above 12 TeV by less than 7%.

The next brightest TeV region is just to the left of MGRO J2019+37 in Figure 2, at Galactic latitude of 80°, and is also coincident with an EGRET source (3EG J2033+4118) and the HEGRA source [13] TeV J2032+413.

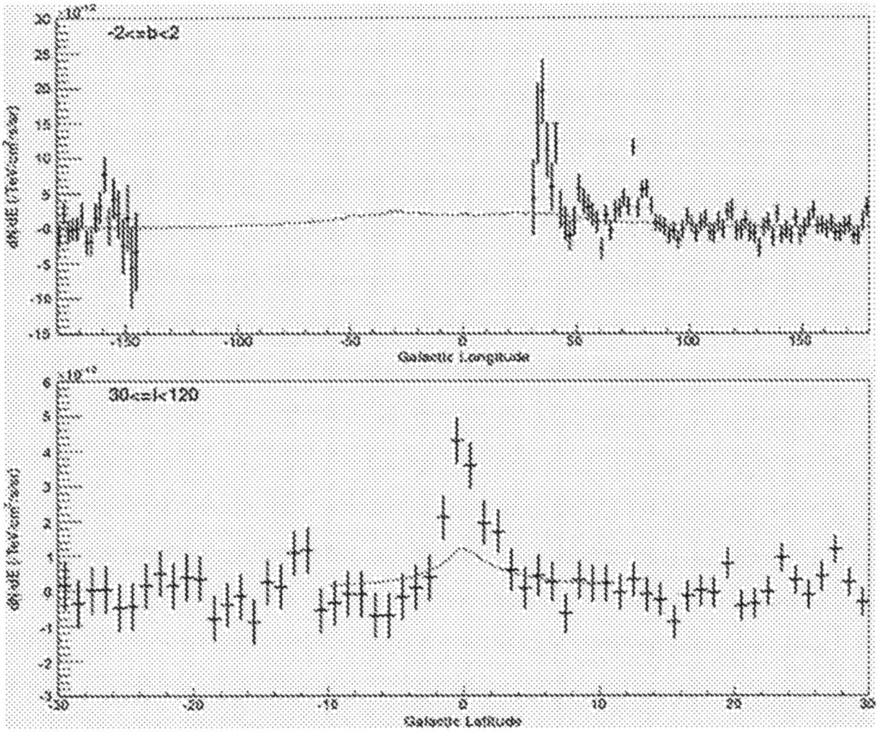

Fig. 3. The top figure is a longitude scan in 2 deg bins of ± 2 deg around the Galactic plane. The lower figure shows a latitude scan of the Galactic plane in 1 degree bins for the inner Galaxy.

The HEGRA source was detected between 1 and 10 TeV with a differential photon spectral index of -1.9± 0.1stat± 0.3sys, which when extrapolated to 12 TeV gives E^2 dN/dE = (7.95± 2.7 stat) x 10^{-13} TeV cm^{-2} sec^{-1} . The Milagro flux in a 3x3 square degree region centered on the HEGRA source at 12 TeV is (2.41± 0.48stat± 072sys) x 10^{-12} cm^{-2} s^{-1} assuming a differential photon source spectrum of $E^{-2.6}$. Thus, the Milagro flux exceeds the HEGRA flux as is expected due to the additional contribution of the diffuse flux in this region. In fact, this region contains the largest matter density as can be seen from the contour lines of Figure 2. Further analysis is required to distinguish what fraction of the gamma rays observed by Milagro is from the HEGRA source or from the diffuse interactions.

Figure 3 shows the profiles in latitude and longitude of the inner Galaxy (from 30-120 degrees in Galactic longitude). The median energy of the

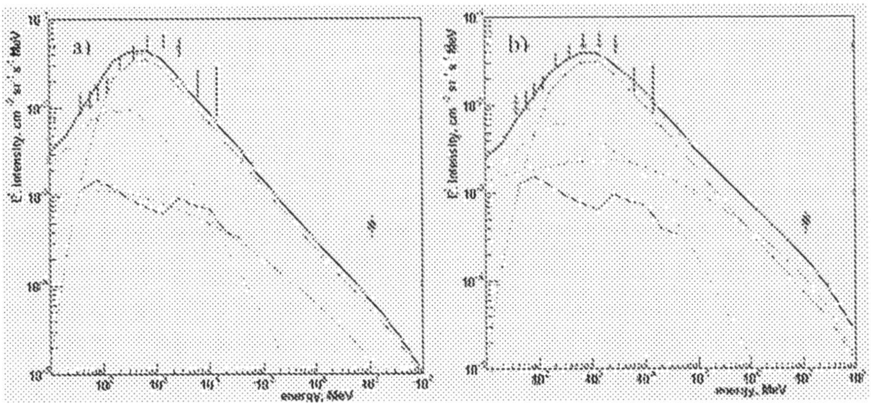

Fig. 4. Gamma-ray spectrum of the diffuse emission from the Cygnus Region of the Galactic Plane. The red bars are the EGRET data and the purple bar is the Milagro measurement with the statistical error shown as a broad line and the systematic error shown as a narrow line. The other lines represent the different components of the emission according to the (a) "conventional" and (b) "optimized" GALPROP model of Strong et. al. [14].

detected gamma rays is $\sim$ 12 TeV. It can be seen that the plane has a width of $\sim \pm 2.5°$ in latitude. The longitude profile shows a large excess in the Cygnus region as well as an increasing flux towards the Galactic center, however with larger error bars because of the limited view of the Galactic center from the latitude of Milagro.

The flux of diffuse emission from the Cygnus Region is measured after subtraction of the contribution from MGRO 2019+37. Using the GALPROP [14] program we estimate the expected flux of diffuse gamma rays. This estimate has contributions from a pion component arising from the interactions of cosmic rays with matter in the region and from gamma rays produced by inverse Compton interactions of cosmic-ray electrons with infrared radiation in the region. We find that the TeV gamma ray flux is about a factor of 5 larger than that predicted by this standard model of cosmic ray production in the Galaxy. These results are shown in figure 4. The same reference [14] gives a model that explains the GeV excess that can also explain the TeV excess however, at energies near 10 TeV it implies that the inverse Compton component is dominant and the inverse Compton component would not show such a strong correlation with the matter density. Other plausible explanations of this observation are that the Cygnus region contains cosmic-ray accelerators thereby increasing the cosmic-ray density in that region relative to the model predictions of GALPROP.

References

1. Atkins, R.. et al. 2005 Phys Rev Lett 95, 251103.
2. Abdo A. et al. 2006, in preparation.
3. Smith, A. et al. 2006, Review, Rapporteur, and Highlight Papers of 29th ICRC in Pune India, 10, 227.
4. Atkins, R.. et al. 2000, ApJ Lett, 533, 119 and Atkins, R.. et al. 2003, ApJ, 583, 824.
5. Atkins, R.. et al. 1999, ApJ Lett, 525, L25.
6. Atkins, R.. et al. 2004, ApJ, 608, 680.
7. Atkins, R.. et al. 2003, ApJ, 595, 803.
8. Atkins, R.. et al. 2005, ApJ, 630, 996 and Atkins, R.. et al., 2004 ApJ Lett, 604, 25.
9. Van der Hucht, K. A., 2001, New Astron. Rev. 45, 135.
10. Bochkarev, N. G. and Sitnik, T. G., 1985, Astrophys. Space Sci. 108, 237.
11. Breen, D. A., 2004, Bull. Astron. Soc. India 32, 325.
12. Roberts, M. S. E., et al., 2002, ApJ. 577, L19.
13. Aharonian, F. et al., 2005, A&A, 431, 197.
14. Strong, A. W., Moskalenko, I.V., and Reimer, O., 2004, ApJ, 613, 962.

OBSERVATION OF GALACTIC SOURCES OF VERY HIGH ENERGY γ-RAYS WITH THE MAGIC TELESCOPE

H. Bartko (for the MAGIC collaboration)

MPI für Physik, Werner-Heisenberg-Institut,
Föhringer Ring 6, D-80805 München, Germany
E-mail: hbartko@mppmu.mpg.de

During its first cycle of observations, the MAGIC (Major Atmospheric Gamma-ray Imaging Cherenkov) telescope has observed very high energy γ-rays from five galactic objects: the Crab Nebula, the SNRs HESS J1813-178 and HESS J1834-087, the Galactic Center and the γ-ray binary LS I +61 303. After a short introduction to the MAGIC telescope and the data analysis procedure, the results of these five sources are reviewed.

Keywords: γ-ray astronomy, Galactic objects, Galactic Center.

1. The MAGIC telescope

MAGIC [1,2] is currently the largest single dish Imaging Air Cherenkov Telescope (IACT) in operation. Located on the Canary Island La Palma (28.8°N, 17.8°W, 2200 m a.s.l.), it has a 17-m diameter tessellated parabolic mirror, supported by a light weight carbon fiber frame. It is equipped with a high quantum efficiency 576-pixel 3.5° field-of-view photomultiplier camera. The analog signals are transported via optical fibers to the trigger electronics and the 300 MSamples/s FADC read-out system, which is currently being upgraded to a 2 GSamples/s FADC system [3].

The physics program of the MAGIC telescope includes both, topics of fundamental physics and astrophysics. In this paper the observations of galactic sources are presented. The observations of extragalactic sources are reviewed elsewhere in these proceedings [4,5].

2. Data analysis

The data analysis was carried out using the standard MAGIC analysis and reconstruction software [6], the first step of which involves the calibration of the raw data [7,8]. It follows the general steps presented in [9,10]: After calibration, image cleaning tail cuts have been applied (see e.g. [11]). The camera images are parameterized by image parameters [12]. The Random Forest method (see [13,14] for a detailed description) was applied for the γ/hadron separation (for a review see e.g. [11]) and the energy estimation.

For each event the arrival direction of the primary γ-ray candidate in sky coordinates is estimated using the DISP-method resulting in VHE γ-ray sky map [15–17]. The angular resolution is $\sim 0.1°$, while source localization in the sky is provided with a precision of $\sim 2'$ [18].

The points of the reconstructed very high energy γ-ray spectrum ($dN_\gamma/(dE_\gamma dAdt)$ vs. true E_γ) are corrected (unfolded) for the instrumental energy resolution [19]. Moreover, a forward unfolding procedure is applied for spectral fits: A candidate spectral law is fitted to the measured data ($dN_\gamma/(dE_\gamma dAdt)$ vs. estimated E_γ) taking the full instrumental energy migration (true E_γ vs. estimated E_γ) into account as described in [20].

The systematic error in the flux level determination is estimated to be 35% the spectral index 0.2, see also [10].

3. Highlights of cycle I

MAGIC's first observation cycle spanned the period from January 2005 to April 2006. About 1/4 of the scientific observation time was devoted to galactic objects. The observations covered included the following types of objects: superova remnants (SNRs), pulsars, pulsar wind nebulae (PWN), microquasar candidates (μQSRs), the Galactic Center (GC), one unidentified TeV source and one cataclysmic variable. In this section the results of the following sources are reviewed: the Crab nebula and some other selected pulsars, the SNRs HESS J1813-178 and HESS J1834-087, the Galactic Center and the γ-ray binary LS I +61 303.

3.1. *The Crab nebula and pulsars*

The Crab nebula is a bright and steady emitter at GeV and TeV energies, what makes it into an excellent calibration candle. This object has been observed extensively in the past over a wide range of wavelengths. Some of the relevant physics phenomena, are expected to happen in the energy domain between 10 and 100 GeV, namely the Inverse Compton peak of the energy sectral energy distribution and the cut-off of the pulsed emission.

Along the first cycle of MAGIC's regular observations, a significant amount of time has been devoted to observe the Crab nebula, both for technical and astrophysical studies. A sample of 12 hours of selected data has been used to measure the energy spectrum down to ~100 GeV, as shown in figure 1 [21]. Also a search for pulsed γ-ray emission from the Crab pulsar has been carried out. The derived upper limits (95% C.L.) are 2.0×10^{-10} ph $s^{-1}cm^{-2}$ at 90 GeV and 1.1×10^{-10} ph $s^{-1}cm^{-2}$ at 150 GeV [22].

Moreover, γ-ray emission was searched for from two milisecond pulsars [23] PSR B1957+20 and PSR J0218+4232, albeit without positive result. The corresponding upper limits ($E_\gamma \geq 115$ GeV) are $F_{\mathrm{PSR\ B1957+20}} \sim 2.3 \times 10^{-11}$ and $F_{\mathrm{PSR\ J0218+4232}} \sim 2.9 \times 10^{-11}$ ph $s^{-1}cm^{-2}$ for the steady emission and $F_{\mathrm{PSR\ J0218+4232}} \sim 6.5\times10^{-12}$ ph $s^{-1}cm^{-2}$ for the pulsed one.

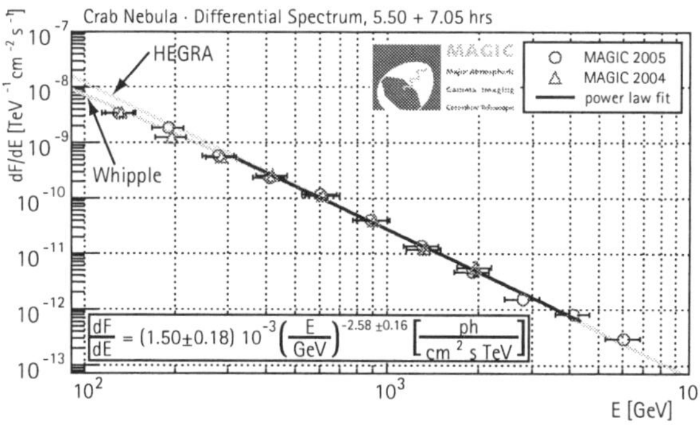

Fig. 1. Energy spectrum above 100 GeV from the Crab nebula measured by MAGIC in two different observation seasons [21].

3.2. *Supernova remnants*

Shocks produced at supernova explosions are assumed to be the source of the galactic component of the cosmic ray flux [24]. In inelastic collisions of high energy cosmic rays with ambient matter γ-rays and neutrinos are produced. These neutral particles give direct information about their source, as their trajectories are not affected by magnetic fields in contrast to the charged cosmic rays. Nevertheless, not all VHE γ-rays from galactic sources are due to the interactions of cosmic rays with ambient matter. There are also other mechanisms for the production of VHE γ-rays like the inverse Compton up-scattering of ambient low energy photons by VHE electrons. For each individual source of VHE γ-rays, the physical processes of particle acceleration and γ-ray emission in this source have to be determined. A powerfull tool is the modelling of the multiwavelength emission of the source and comparison to multiwavelength data.

Within its program of observation of galactic sources, MAGIC has taken data on a number of supernova remnants. VHE γ-ray emission from the SNRs HESS J1813-178 [9] and HESS J1834-087 (W41) [25] has been observed. These results have confirmed SNRs as a well established population of VHE γ-ray emitters.

The energy spectrum of HESS J1813-178, which is spatially coincident with SNR G12.82-0.02, can be fitted with a hard-slope power law, described as $dN_\gamma/(dAdtdE) = (3.3 \pm 0.5) \times 10^{-12}(E/\mathrm{TeV})^{-2.1\pm0.2}\ \mathrm{cm^{-2}s^{-1}TeV^{-1}}$. The source HESS J1834-087 is spatially coincident with SNR G23.3-0.3 (W41). The observed differential γ-ray flux is consistent with a power law $dN_\gamma/(dAdtdE) = (3.7 \pm 0.6) \times 10^{-12}(E/\mathrm{TeV})^{-2.5\pm0.2}\ \mathrm{cm^{-2}s^{-1}TeV^{-1}}$. A source extension of $(0.14 \pm 0.04)^\circ$ is derived. A spatial superposition of the γ-ray source with a massive molecular cloud observed by its ^{13}CO and ^{12}CO emission was found. Although the mechanism responsible for the VHE radiation remains yet to be clarified, this is a hint that it could be produced by high energy hadrons interacting with the molecular cloud.

3.3. *Galactic center*

The Galactic Center region contains many remarkable objects which may be responsible for high-energy processes generating γ-rays like a supermassive black hole, supernova remnants, candidate pulsar wind nebulae, a high density of cosmic rays, hot gas and large magnetic fields. Moreover, the Galactic Center may emit the highest VHE γ-ray flux from the annihilation of possible dark matter particles [26] of all proposed dark matter particle annihilation sources.

One motivation for this observations was the possibility to indirectly detect dark matter through its annihilation into VHE γ-rays, see e.g. [26].

The Galactic Center was observed with the MAGIC telescope [10] under large zenith angles resulting in the detection of a differential γ-ray flux, consistent with a steady, hard-slope power law between 500 GeV and about 20 TeV, described as $dN_\gamma/(dAdtdE) = (2.9 \pm 0.6) \times 10^{-12}(E/\mathrm{TeV})^{-2.2\pm0.2}\ \mathrm{cm^{-2}s^{-1}TeV^{-1}}$. This result confirms the previous measurements by the HESS collaboration. The VHE γ-ray emission does not show any significant time variability; the MAGIC measurements rather affirm a steady emission of γ-rays from the GC region on time scales of up to one year.

The VHE γ-ray source is centered at (RA, Dec)=($17^h45^m20^s$, $-29^\circ 2'$). The excess is point-like, it's location is consistent with SgrA*, the candidate PWN G359.95-0.04 as well as SgrA East.

The nature of the source of the VHE γ-rays has not yet been identi-

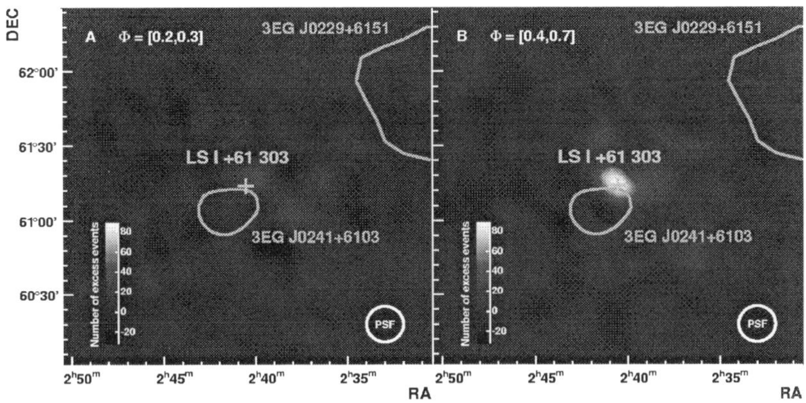

Fig. 2. Smoothed maps of γ-ray excess events above 400 GeV around LS I +61 303, for observations around periastron (A) and latter orbital phases (B) [27].

fied. The power law spectrum up to about 20 TeV disfavours dark matter annihilation as the main origin of the detected flux. The absence of flux variation indicates that the VHE γ-rays are rather produced in a steady object such as a SNR or a PWN, and not in the central black hole.

3.4. *The γ-ray binary LS I +61 303*

This γ-ray binary system is composed of a B0 main sequence star with a circumstellar disc, i.e. a Be star, located at a distance of $\sim$2 kpc. A compact object of unknown nature (neutron star or black hole) is orbiting around it, in a highly eccentric ($e = 0.72 \pm 0.15$) orbit.

LS I +61 303 was observed with MAGIC for 54 hours between October 2005 and March 2006 [27]. The reconstructed γ-ray map is shown in figure 2. The data were first divided into two different samples, around periastron passage (0.2-0.3) and at higher (0.4-0.7) orbital phases. No significant excess in the number of γ-ray events is detected around periastron passage, whereas there is a clear detection (9.4σ statistical significance) at later orbital phases. Two different scenarios have been involved to explain this high energy emissions: the microquasar scenario where the γ-rays are produced in a radio-emitting jet; or the pulsar binary scenario, where they are produced in the shock which is generated by the interaction of a pulsar wind and the wind of the massive companion. See [28] for more details.

Acknowledgements. We thank the IAC for the excellent working conditions at the ORM in La Palma. The support of the German BMBF and MPG, the Italian INFN, the Spanish CICYT is gratefully acknowledged. This work was also supported by ETH research grant TH-34/04-3, and the Polish MNiI grant 1P03D01028.

References

1. Baixeras, C. et al., 2004, NIM, A518, 188.
2. Cortina, J. et al. (MAGIC Collab.), 2005, Proc. of the 29th ICRC, Pune, India, 5-359, astro-ph/0508274.
3. Bartko, H. et al., 2005, Nucl. Instrum. Meth., A548, 464.
4. Errando, M., These proceedings.
5. Garczarczyk, M., These proceedings.
6. Bretz, T. & R. Wagner (MAGIC Collab.), 2003, Proc. of the 28th ICRC, Tsukuba, Japan, 2947.
7. Gaug, M. et al. (MAGIC Collab.), 2005, Proc. of the 29th ICRC, Pune, India, 5-375, astro-ph/0508274.
8. Bartko, H. et al. (MAGIC Collab.), 2005, astro-ph/0505165.
9. Albert, J. et al., 2006, ApJ, 637, L41.
10. Albert, J. et al., 2006, ApJ, 638, L101.
11. Fegan, D. J., 1997, J Phys G, 23, 1013.
12. Hillas, A. M., 1985, Proc. of the 19th ICRC, La Jolla, 3, 445.
13. Bock, R. K. et al., 2004, NIM, A516, 511.
14. Breiman, L., 2001, Machine Learning, 45, 5.
15. Fomin, V. P. et al., 1994, Astroparticle Physics, 2, 137.
16. Lessard, R. W. et al., 2001, Astroparticle Physics, 15, 1.
17. Domingo-Santamaria, E. et al. (MAGIC Collab.), 2005, Proc. of the 29th ICRC, Pune, India, 5-363, astro-ph/0508274.
18. Riegel, B. et al. (MAGIC Collab.), 2005, Proc. of the 29th ICRC, Pune, India, 5-219, astro-ph/0508274.
19. Anykeev, V. B., Spiridonov, A. A. & Zhigunov, V.B., 1991, Nucl. Instrum. Meth., A303, 350.
20. Mizobuchi, S. et al. (MAGIC Collab.), 2005, Proc. of the 29th ICRC, Pune, India, 5-323, astro-ph/0508274.
21. Wagner, R. et al. (MAGIC Collab.), 2005, Proc. of the 29th ICRC, Pune, India, 4-163, astro-ph/0508244.
22. Lopez, M. et al. (MAGIC Collab.), 2005, Proc. of the 29th ICRC, Pune, India, 4-243, astro-ph/0508244.
23. Oña-Wilhelmi, E. et al. (MAGIC Collab.), 2005, Proc. of the 29th ICRC, Pune, India, 4-247, astro-ph/0508244.
24. Baade, W. & Zwicky, F., 1934, Phys. Rev., 46, 76.
25. Albert, J. et al. (MAGIC Collab.), 2006c, ApJ, 643, L53.
26. Bartko, H. et al. (MAGIC Collab.), Proc. of the 29th ICRC, Pune, India, 4-17, astro-ph/0508273.
27. Albert, J. et al. (MAGIC Collab.), 2006, Science 312, 1771.
28. Sidro, N. (MAGIC Collab.), 2006, astro-ph/0610925.

OBSERVATION OF EXTRAGALACTIC SOURCES OF VERY HIGH ENERGY γ-RAYS WITH THE MAGIC TELESCOPE

M. Errando for the MAGIC Collaboration*

Institut de Fsica d'Altes Energies (IFAE)
Edifici Cn., Universitat Autnoma de Barcelona
08193 Bellaterra (Spain)
E-mail: errando@ifae.es

MAGIC is currently the largest single dish ground–based imaging air Cherenkov telescope in operation. During its first cycle of observations more than 20 extragalactic objects have been observed, and very high energy γ-ray signals have been detected in several of them. The results of this observations are presented, together with a discussion of the spectral characteristics and the flux variability of the detected sources.

1. The MAGIC telescope

MAGIC is a 17m imaging atmospheric Cherenkov telescope for Very High Energy (VHE) γ-ray observations. It is located in the Roque de los Muchachos Observatory on the canary island of La Palma. It can explore gamma-rays at energies below 100 GeV, which is critical to observe sources at medium redshifts where absorption of gamma photons by the Extragalactic Background Light (EBL) attenuates the emission at higher gamma energies.

The detection and characterization of VHE γ-ray emitting Active Galactic Nuclei (AGN) is one of the main goals for ground–based γ-ray astronomy. This studies open the possibility of exploring the physics of the relativistic jets in AGN, relate the flux of photons in different energy bands (optic, X-rays and γ-rays), perform population studies of blazar objects and even extract information about the EBL photon density. MAGIC can also be repositioned in few seconds to do observations of Gamma-Ray Bursts (GRBs) in their prompt phase of early afterglow, although none of these has resulted in a positive detection to date. GRB observations done by

*Updated collaborator list at http://wwmagic.mppmu.mpg.de/collaboration/members

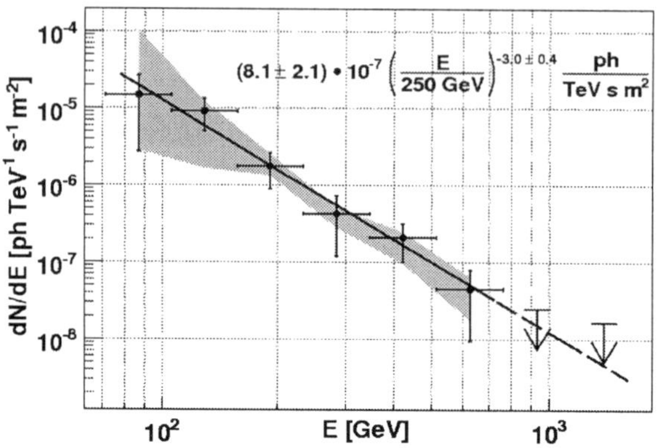

Fig. 1. Differential energy spectrum of 1ES 1218+30.4.

MAGIC are reviewed elsewhere in these proceedings,[1] as well as the results of galactic observations.[2]

2. Cycle I observations

The first cycle of observations of the MAGIC telescope extended from January 2005 to April 2006. The observations of extragalactic objects covered High-frequency peaked BL Lacs (HBLs), selected flat spectrum radio quasars and low-frequency peaked BL Lacs located at low redshifts, known TeV-emitting HBLs, the Ultra–Luminous infrared Galaxy (ULIRG) Arp 220 and the radio galaxy M87. In this section, the following results are reviewed: the recently discovered VHE γ-ray emission from the HBLs 1ES 1218+30.4, Mkn 180 and PG 1553+113, an upper limit on the VHE emission of Arp 220 and the observations of the known TeV blazars Mkn 501, Mkn 421, 1ES 1959+650 and 1ES 2344+514.

2.1. *Observation of VHE γ-ray candidate sources*

Selection of VHE γ-ray emitting candidate sources follows criteria based on the spectral properties of the considered objects. Using both Synchrotron Self–Compton (SSC) and hadronic models, the spectral energy distribution of the candidate AGN can be extrapolated to MAGIC energies to predict its observability. The preferred candidates are usually strong X-ray emitters, but selections based on the optical band are also considered. Moreover, other non–blazar objects as the giant radio–galaxy M87 or the ULIRG Arp 220 have also been observed, although none of these observations re-

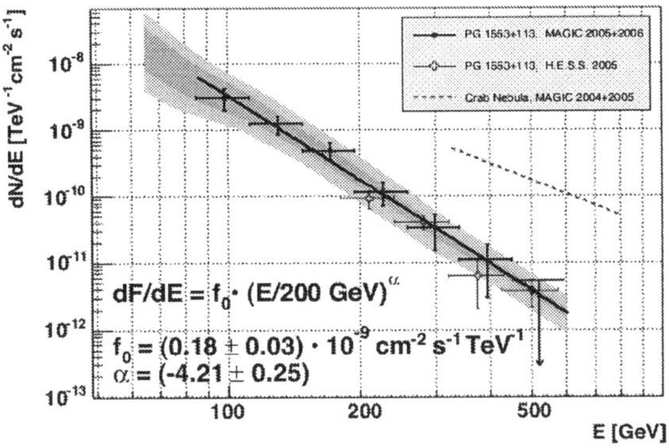

Fig. 2. Differential energy spectrum of PG 1553+113.

sulted in a positive detection to date.

1ES 1218+30.4[3] is the first source discovered by MAGIC and one of the most distant VHE γ-ray sources. This HBL located at redshift $z = 0.182$ was previously observed by Whipple and HEGRA, but only upper limits where derived. MAGIC observed it during 8.2 h in January 2005, obtaining a γ-ray signal of 6.4 σ significance in the 87 to 630 GeV energy range. The observed differential energy spectrum can be fitted by a simple power law with photon index 3.0 ± 0.4, and no time variability on timescales of days was found within statistical errors.

PG 1553+113[4] is a distant BL Lac of undetermined redshift that was observed by the MAGIC telescope in 2005 and 2006, and has also been recently detected by the H.E.S.S. collaboration.[5] A VHE γ-ray signal has been detected by MAGIC with an overall significance of 8.8 σ, showing no significant flux variations a daily timescale. However, the flux observed in 2005 was significantly higher compared to 2006. The differential energy spectrum between 90 and 500 GeV can be well described by a power law with photon index 4.2 ± 0.3, being steeper than that of any other known BL Lac object. This spectrum can be used to derive an upper limit on the source redshift. Assuming an EBL model by Kneiske et al.[6] and a physical limit on the intrinsic source spectrum (intrinsic photon index $\alpha_{int} < -1.5$),[7] an upper limit on the source redshift of $z < 0.78$ has been derived.

Mkn 180[8] is an HBL that had an optical outburst in March 2006

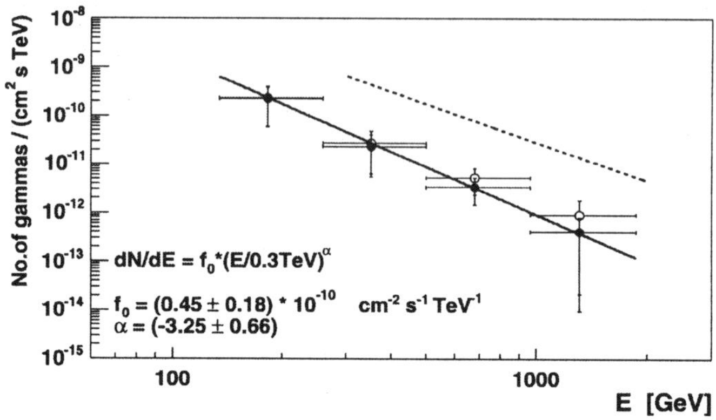

Fig. 3. Differential energy spectrum of Mkn 180. The filled circles show the measured spectrum while the open circles display the intrinsic spectrum of the source where the EBL absorption has been removed.

observed by the KVA 35 cm telescope also located at the Roque de los Muchachos Observatory. It triggered the observations with the MAGIC telescope in the GeV–TeV band, resulting on the first detection of VHE γ-ray emission from this source. A total of 12.4 h of data were recorded during eight nights, giving a 5.5 σ significance detection. The integral flux above 200 GeV corresponded to 11% of the Crab Nebula flux, and the differential energy spectrum could be fitted by a power law with a photon index of 3.3 ± 0.7.

Arp 220[9] is the nearest ULIRG (located at about 72 Mpc) and the one with the largest supernova explosion rate (4 ± 2 per year), and therefore is a good VHE γ-ray emitting candidate. With 15.5 h of data taken by MAGIC upper limits in the 0.16-1.3 TeV band were imposed, which are compatible with a complete multiwavelength modeling of the source.[10]

3. Monitor of known TeV blazars

The improved sensitivity and energy threshold of MAGIC with respect to the former generation of γ-ray telescopes allows a detailed study of the spectral features and flux variations of known TeV emitters.

Mkn 421[11] is the closest TeV blazar ($z = 0.031$) and the first extragalactic VHE source detected with a ground–based γ-ray telescope.[12] MAGIC has observed this source between November 2004 and April 2005 obtaining 25.6 h of data, and including 1.5 h of simultaneous observations with the H.E.S.S. array.[13] Integral flux variations up to a factor of four are observed between different observation nights, although no significant intra–night variations have been recorded despite the high sensitivity of the MAGIC telescope for this kind of search. This flux variability showed a clear correlation with between γ-ray and X-ray fluxes, favoring leptonic emission models. The energy spectrum between 100 GeV and 3 TeV shows a clear curvature. After correcting the measured spectrum for the effect of γ-attenuation caused by the EBL light assuming a model of Primack et al.,[14] there is an indication of an inverse Compton peak around 100 GeV.

1ES 2344+514 was first detected in 1995 by Whipple when it was in flaring state,[15] and HEGRA reported later a weak detection in quiescent state.[16] MAGIC obtained a VHE γ-ray signal with 11.0 σ significance from 23.1 h of data. The source was in quiescent state during the observations, with a flux level compatible with the HEGRA results, but showing a softer spectrum. A detailed publication on the analysis and results of these observations is in preparation.

1ES 1959+650[17] presented in 2002 a VHE γ-ray flare without any counterpart in X-rays.[18] This behavior can not be easily explained by the SSC mechanism in relativistic jets that successfully explain most of the VHE γ-ray production in other HBLs. MAGIC observed this object during 6 h in 2004, when it was in low activity both in optical and X-ray bands, detecting a γ-ray signal with 8.2 σ significance. The differential energy spectrum between 180 GeV and 2 TeV can be fitted with a power law of photon index 2.72 ± 0.14, which is consistent with the slightly steeper spectrum seen by HEGRA at higher energies,[19] also during periods of low X-ray activity.

Mkn 501 is a close TeV blazar, first detected by Whipple in 1996.[20] MAGIC observed it during 55 h in 2005, including 34 h in moderate moonlight conditions. The source was in low state (30-50% of the Crab Nebula flux for $E > 200$ GeV) during most of the observation time but showed two episodes of fast and intense flux variability, with doubling times of about 5 minutes. Changes in the spectral slope with the flux level have been observed for the first time in timescales of 10 minutes. A detailed publication on the analysis and results of these observations is in preparation.

4. The MAGIC II telescope

The MAGIC collaboration is currently constructing a second telescope in the same site at the Roque de los Muchachos Observatory, which will operate in stereo mode with the MAGIC telescope improving the overall sensitivity. The MAGIC II telescope is a clone of the existing one with two main improvements: a fine pixelized camera with cluster design that will allow an update of the photomultipliers to hybrid photon detectors once this technology is ready to be used, and an ultra–fast signal readout that will sample the pulses coming from each pixel at 2.5 GHz, allowing a better signal reconstruction and the use of timing analysis techniques. The commissioning phase of MAGIC II will start at the end of 2007.

5. Conclusions

MAGIC has concluded its first cycle of observations having detected seven extragalactic VHE γ-ray sources, including three objects never detected before at these wavelengths. The high sensitivity and low energy threshold allowed detailed studies of the spectral features of these sources, as well as the observation of flux variability in short timescales.

References

1. Garczarczyk M., these proceedings.
2. Bartko H., these proceedings.
3. Albert J. et al., ApJ Letters 642 (2006) 119.
4. Albert J. et al., ApJ Letters, in press.
5. Aharonian F. et al., A&A Letters 448 (2006) 19.
6. Kneiske T.M., Bretz T., Mannheim K. & Hartmann D.H., A&A 413 (2004) 807.
7. Aharonian F. et al., Nature 440 (2006) 1018.
8. Albert J. et al., ApJ Letters 648 (2006) 105.
9. Albert J. et al., ApJ, in press.
10. Torres D.F., ApJ 617 (2004) 966.
11. Albert J. et al., submitted to ApJ in March 2006.
12. Punch M. et al., Nature 358 (1992) 477.
13. Mazin D. et al., Proc. 29th ICRC (2005) 4 331.
14. Primack J. et al., AIP Conf. Proc. 745 (2005) 23.
15. Catanese M. et al., ApJ 501 (1998) 616.
16. Tluczycont M. et al, Proc. 28th ICRC (2003) 5 2447.
17. Albert J. et al., ApJ 639 (2006) 761.
18. Krawczynski H. et al, ApJ 601 (2004) 151.
19. Aharonian F. et al., A&A Letters 406 (2003) 9.
20. Quinn J. et al., ApJ Letters 456 (1996) 83.

INITIAL STEREO ANALYSIS OF MRK 421 FROM THE VERITAS TELESCOPES

S. B. HUGHES* for the VERITAS Collaboration

Department of Physics, Washington University,
1 Brookings Drive, CB 1105,
St. Louis, Missouri 63130, USA
**E-mail: shughes78@gmail.com*

VERITAS (Very Energetic Radiation Imaging Telescope Array System), an array of ground-based gamma-ray telescopes in southern Arizona, USA, has been taking data in hardware stereo mode since March, 2006. The April–May 2006 dark run provided a large set of data from two telescopes on the known blazar Markarian (Mrk) 421. An initial analysis produced a light curve and preliminary cuts showing the two telescope array's angular resolution to be 0.19°. The remaining two VERITAS telescopes will be brought online by January, 2007.

Keywords: gamma-ray astronomy; blazars; Cherenkov telescopes.

1. Introduction

Some of the most powerful and exciting sources in the night sky are Active Galactic Nuclei (AGN). These compact sources, thought to be powered by mass-accreting black holes, eject matter along the poles while accelerating it to very high energies. A blazar is an AGN that has one of its jets directed towards the Earth, allowing observers to view straight down the stream of relativistic matter. Blazars show X-ray and gamma-ray flux variability on time scales of minutes. Spectral Energy Distributions (SEDs) from blazars show two distinct peaks: one in X-rays, the other in TeV gamma rays. These peaks are attributed to Synchrotron and inverse-Compton emission of a single population of high energy electrons, respectively.

To observe the TeV gamma rays emitted by blazars, it is necessary to use ground-based observatories which detect Cherenkov light from air showers produced when gamma rays interact with the Earth's atmosphere. Stand-alone gamma-ray telescopes such as the Whipple 10 m and CAT have been successful at studying blazars since 1992, when Mrk 421 was initially detected as a gamma-ray source.[1] The future brings arrays of these telescopes operating in stereo, such as those of the H.E.S.S.[2] and CANGAROO[3] collaborations. Requiring multiple telescope coincidence means greater background rejection and increased sensitivity over a small field of view. VERITAS is the next of these telescope arrays to come fully online, although it was the first such project to be proposed.

2. The VERITAS Telescopes

VERITAS is an array of imaging atmospheric Cherenkov telescopes (IACTs) currently being constructed at the Whipple Observatory Base Camp, at an altitude of 1275 m on Mt. Hopkins in southern Arizona, USA.[4] Originally proposed as an array of seven 12 m optical reflectors in 1996, final budget approval in the U.S. came in 2003 for the building of four telescopes. The current site is only temporary. It is planned to move the experiment to its permanent location on Kitt Peak after two years of data taking.

Built as a follow-up to the Whipple 10 m gamma-ray telescope, VERITAS is designed to detect 100 GeV–50 TeV gamma rays. VERITAS has a better single photon angular resolution, higher reduction of background events, and achieves a better sensitivity than the Whipple telescope.

Fig. 1. The camera for each VERITAS telescope consists of 499 PMT pixels with a 3.5° field of view. The camera housing is 1.8 m across, to accommodate future expansions and upgrades.

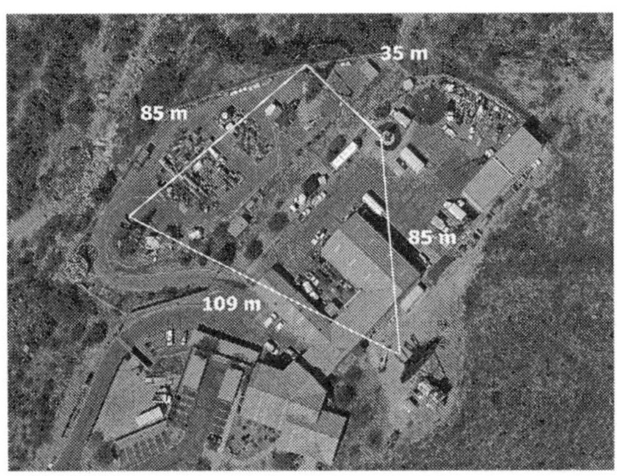

Fig. 2. The current layout for the four VERITAS telescopes at the Whipple Base Camp.

Each VERITAS telescope has a camera consisting of 499 photomultiplier tube (PMT) pixels, each with an angular diameter of 0.15°. They are located in a spacious housing 1.8 m across (see Fig. 1), in anticipation of future camera expansions and upgrades (for instance, the Whipple camera was upgraded three times in the past ten years). The current camera has a field of view of 3.5°, similar to that of the Whipple telescope.

The data acquisition system for VERITAS uses a 500 MHz flash analog to digital converter (FADC) system developed at Washington University.[5] Each channel has 32 μs of lookback memory. The FADC system reduces the night sky background noise in images as the signal can be extracted with a narrow ($\sim$ 6 ns) readout window. There are also separate high- and low-gain paths through the acquisition system. If a channel's input exceeds the high gain limit, a bit is flipped and the signal travels through a separate low gain path, being delayed in memory by a fixed amount, where it can be read out without saturation.

VERITAS also has a true array trigger for data acquisition. PMT pulses are continuously saved into memory by the FADCs while potential triggers pass up the trigger chain. Only when an array trigger is confirmed does data-taking stop to read out the necessary information from the flash memory.

The original layout of the four VERITAS telescopes was a "Mercedes star" pattern, with one telescope at the center and the other three spaced equidistantly around a circle of radius 80 m from the central telescope. The temporary construction at the Whipple Base Camp forced a change in plan, as seen in Figure 2. The result is two pairs of telescopes of 85 m spacing, with the pairs between 35 m and 109 m apart. All four telescopes are able to work together as one, or separately in any desirable grouping.

The first of the VERITAS telescopes (T1) was operated as a prototype with half of the PMTs and one third of its mirrors at the Whipple Base Camp starting in 2004.[6] It became fully operational in February, 2005. The second telescope (T2) saw first light in September, 2005. The two telescopes were operated in unison towards the end of the year producing software stereo data until March, 2006 when true, hardware stereo mode operation began. T3 and T4 are nearing completion, with the anticipation of the full four-telescope array being operational by January, 2007.

3. Stereo imaging of air showers

The Earth's atmosphere is opaque to high energy particles such as gamma rays. However, they can interact with the atmosphere and pair produce, initiating a cascading air shower that emits Cherenkov photons whose peak emission is in the UV/blue. The shower light spreads out on the ground over an area of $\sim 10^5$ m^2. IACTs take advantage of this, resulting in a large effective area with relatively small telescopes.

Stereo imaging of these showers, where detection by more than one telescope is required to record an event, results in much higher sensitivity due to increased rejection of background events. Gamma ray initiated air showers produce a uniform Cherenkov photon density at the ground, while cosmic ray initiated air showers do not. Hence, the latter will not trigger the entire telescope array as often. Local muons, the dominant background in single Cherenkov telescopes, are also strongly suppressed at the trigger level. The use of multiple telescopes results in a single gamma ray angular resolution of about 0.1° and energy resolution of 20%.

Data for this analysis were taken using two different observing modes. The difference between the two is the location and shape of the source (ON) and background (OFF) regions of the camera (the regions always have equal areas). The first is Tracking mode, where the source is located in the center of the camera. The ON region is then a circle at the camera's center, while the OFF region is an annulus of the same area located outside the ON region, with enough space in between to prevent cross contamination of the regions. The second observing mode is Wobble, where the source is offset from the camera center by a fixed amount (here, 0.3°). The ON region is centered on the source, and the OFF region is then of identical size and shape, offset by the same amount in the opposite direction. Most of the data for this analysis were taken in Tracking mode.

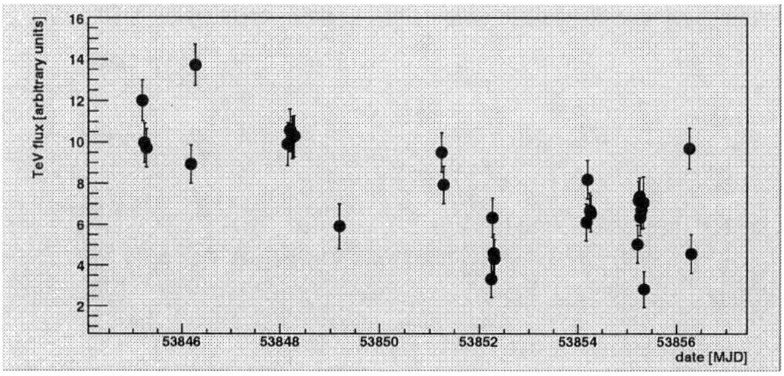

Fig. 3. Light curve for Mrk 421 data taken by VERITAS in April–May 2006. Each data run is represented by one point on the graph. Error bars are on the one sigma confidence level.

4. Data and results

During the April–May 2006 dark run, VERITAS took a significant amount of data on the nearby blazar Mrk 421, to be used for calibration and performance testing purposes. These data consisted of 14.3 hrs of tracking and wobble data, taken on nine separate nights. Data were analyzed using the *eventdisplay* package developed at the University of Leeds by J. Holder and

G. Maier, which showed signal from Mrk 421 with a total statistical significance of 32 sigma. The absolute flux calibration has not yet been finalized. Figure 3 shows a preliminary light curve from these observations, illustrating both the variable nature of Mrk 421 and the ability of VERITAS to detect this variability on short time scales.

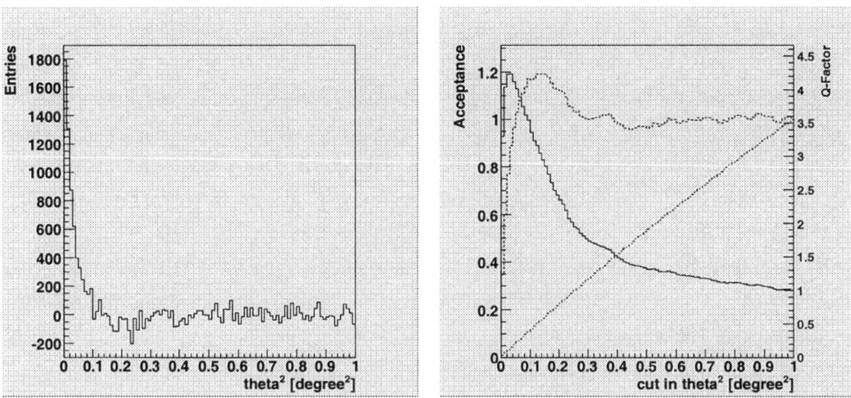

Fig. 4. Plots of θ^2 for the Mrk 421 data. (a) The resulting source distribution after subtracting data in the OFF region from data in the ON region. (b) Calculating the Q-factor for θ^2, showing Q (solid line), signal acceptance (dashed line) and background acceptance (dot-dashed line).

Initial analysis of the data focused on θ^2, the square of the angular distance of the reconstructed arrival directions of the primary gamma rays from the source direction. The distribution resulting from subtracting data in the OFF region from that in the ON region is shown in Figure 4a. To determine the optimal angular cut, we calculated the Q-factor of the θ^2 distribution (see Fig. 4b).[a] This resulted in a maximum Q-factor of 3.08 at a value of $\theta^2 = 0.035$. This corresponds to a signal acceptance of 66% with 95% background rejection and an angular resolution of 0.19°.

5. Future observations and conclusions

In conclusion, the first two VERITAS telescopes are performing as expected, and the full four-telescope array will be coming online by the start of 2007. The complete system will allow for many detailed studies of blazars as well

[a]The Q-factor of a cut is defined by its gamma-ray acceptance divided by the square root of the background acceptance. The Q-factor scales with the statistical significance of weak signals obtained with a certain cut.

as the detection of new sources. Combining flux measurements and spectral correlations obtained with data in different energy bands from space-borne X-ray and gamma-ray telescopes will help break model degeneracies and permit measurements of various jet parameters, as well as allow us to study jets at their base (radius < 1 pc).

With the scheduled launch of GLAST (Gamma-ray Large Area Space Telescope) in 2007, and the simultaneous operation of the H.E.S.S., VERITAS, and MAGIC experiments, we are at a very exciting time in high energy astrophysics, where it will be possible to monitor blazars simultaneously at energies from MeV to TeV. Observing blazars with radio, IR, and optical telescopes, in X-rays with INTEGRAL and Suzaku, and in gamma rays with GLAST and Cherenkov telescopes will allow us to test the models with unprecedented spectral coverage and sensitivity. We hope that the observations will make it possible to identify unambiguously the nature of the accelerated particles (protons or electrons/positrons). Once the emission mechanism is interpreted, the observations will give information about the jet parameters (magnetic field, particle to magnetic field energy density), and thus contribute to our understanding of the structure of AGN jets.

Acknowledgments

SBH would like to acknowledge support from the American Astronomical Society and the National Science Foundation in the form of an International Travel Grant making travel to Erice possible.

VERITAS is supported by grants from the U. S. Department of Energy, the U. S. National Science Foundation, the Smithsonian Institution, by NSERC in Canada, by Science Foundation Ireland, and by PPARC in the U. K. It is being built through a collaboration of nine primary universities and the Smithsonian Astrophysical Observatory, as well as several other contributing institutions.

References

1. M. Punch *et al.*, *Nature* **358**, 477 (August 1992).
2. F. Aharonian *et al.*, *Astronomy & Astrophysics* **457**, 899 (October 2006).
3. R. Enomoto *et al.*, *ApJ* **638**, 397 (February 2006).
4. T. C. Weekes *et al.*, *Astroparticle Physics* **17**, 221 (May 2002).
5. J. Buckley, "High Speed Electronics for Atmospheric Cherenkov Detectors", in *International Cosmic Ray Conference*, 1999.
6. J. Holder *et al.*, *Astroparticle Physics* **25**, 391 (July 2006).

THE GLAST MISSION AND OBSERVABILITY OF SUPERNOVAE REMNANTS

OMAR TIBOLLA* ON BEHALF OF THE GLAST-LAT COLLABORATION

Dipartimento di Fisica G.Galilei, Padova University,
Via F.Marzolo 8, 35131 Padova, Italy
** E-mail: omar.tibolla@pd.infn.it*

The Gamma-ray Large Area Space Telescope (GLAST) is a collaboration of several countries: France, Germany, Italy, Japan, Sweden and United States of America. GLAST is a satellite-based observatory that will study the Cosmos in the Energy Range 10 keV - 300 GeV and consists of two different instruments: the Large Area Telescope (LAT) and the GLAST Burst Monitor (GBM). The two instruments are ready and now are being tested and integrated with the spacecraft: the launch is scheduled for October 2007. GLAST will improve the knowledge about several astrophysical sources (AGNs, Pulsars, GRBs, etc.). In this paper we will consider the scientific case of gamma-ray emitting Supernovae Remnants (SNRs).

Keywords: GLAST; gamma-astronomy; Supernovae Remnants; NASA.

1. Introduction

GLAST [1] is a satellite-based observatory consisting of two different instruments (LAT and GBM) that will provide for correlative measurements of gamma-ray phenomena in the energy range between 10 keV and 300 GeV. Concerning the GLAST hardware status, the LAT final assembly is complete, the GBM instruments are ready and the collaboration is currently testing the two instruments and integrating them with the spacecraft. The observatory is scheduled to be launched in October 2007 and will provide improvements on several fields of high energy astrophysics.

2. The LAT

The main instrument of GLAST is the Large-Area Telescope [2]: a pair conversion telescope, which takes much of its basic design concepts from its predecessor EGRET (Energetic Gamma Ray Experiment Telescope) [3],

which flew onboard the Compton Gamma Ray Observatory (CGRO) [4] and took data between 1991 and 2000.
LAT is composed by in three different parts:

- an Anticoincidence Shield (ACD),
- a Tracker (TKR),
- a Calorimeter (CAL).

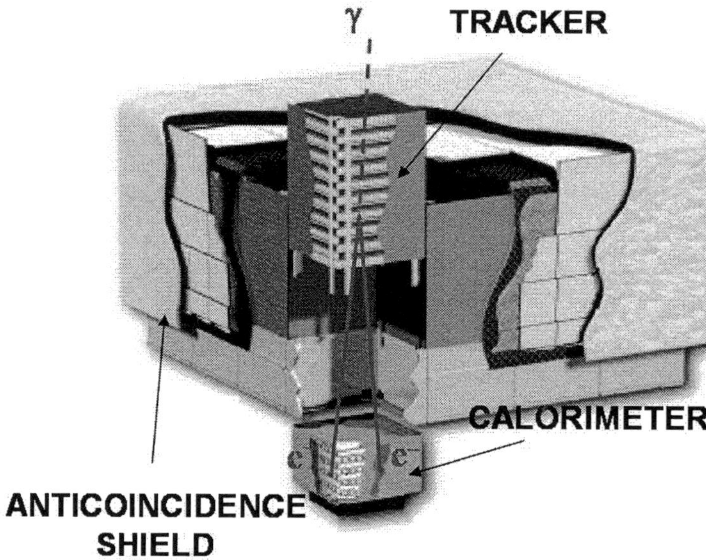

Fig. 1. The LAT is formed by 16 modules (or tower), each tower contains a Traker and a Calorimeter; the modules are surrounded by an anticoincidence shield.

ACD consists of overlapping tiles made of plastic scintillator (Bicron-408) sensitive to charged particles and read out by optical fibers and phototubes; ACD is segmented in order to avoid self veto from Calorimeter backsplash (this caused 50% of loss in EGRET Effective Area) and also for micrometeorites. ACD envelopes the 16 identical towers and each tower is divided in two parts: the upper part is the TKR, the lower the CAL. The TKR is based on silicon-strip technology and is formed by 18 layers in each tower; each layer consists of two orthogonal Single-Slide Detector (SSD) planes of silicon microstrip sensors for detecting the particle tracks and of a thin tungsten-alloy conversion foil. In each tower, below the TKR, is situated a Calorimeter made of hodoscopic stacks of thallium-doped cesium iodide crystals, where the energy deposited by EM shower is converted into a light signal. A schematic rapresentation of the LAT is shown in Figure 1.

LAT was designed to improve on the performance of EGRET, in terms of field-of-view, resolution, dead-time, background rejection and hence sensitivity; all in all GLAST is designed to provide the 3rd EGRET Catalog (i.e. 9 years of observation) in one day. Table 1 compares some of the performances of EGRET with that expected of the LAT; Figure 2 shows some instrument response functions (IRFs) of LAT; more detailed LAT specifications and expected performances can be found in [5].

Table 1. Comparing the performances of EGRET and LAT.

	EGRET	LAT	Factor of improvement
Energy	20 MeV - 30 GeV	20 MeV - 300 GeV	10
Energy Resolution	~10%	~10%	1
Peak Effective Area	1500 cm^2	10000 cm^2	6
Field of View	0.5 sr	2.4 sr	4
Sensitivity (1 year)	$\sim 10^{-7}\gamma\ cm^{-2}s^{-1}$	$3 \times 10^{-9}\gamma\ cm^{-2}s^{-1}$	30
Source Localization	15'	<0.5'	30
Dead-time	100 ms	$< 50\ \mu s$	2000

3. The GBM

The GLAST Burst Monitor [6] is similar in design to another highly successful instrument onboard the CGRO: the Burst And Transient Source Experiment (BATSE). GBM includes two sets of detectors:

- 12 sodium iodide (NaI) scintillators for lower energies;
- 2 cylindrical bismuth germanate (BGO) scintillators for higher energies.

The NaI detectors are sensitive in the lower end of the energy range, from a few of keV to about 1 MeV, and provide burst triggers and locations. The BGO detectors cover the energy range from ~150 keV to ~30 MeV, providing a good overlap with the NaI scintillators at the lower end and with the LAT at the higher end. GBM has a sky coverage of 8 steradians, which enables prompt alerts to the LAT detector and hence to the ground-based gamma-ray telescopes; the precision of localization is expected to be better than 15° in the first several seconds and the ultimate resolution is

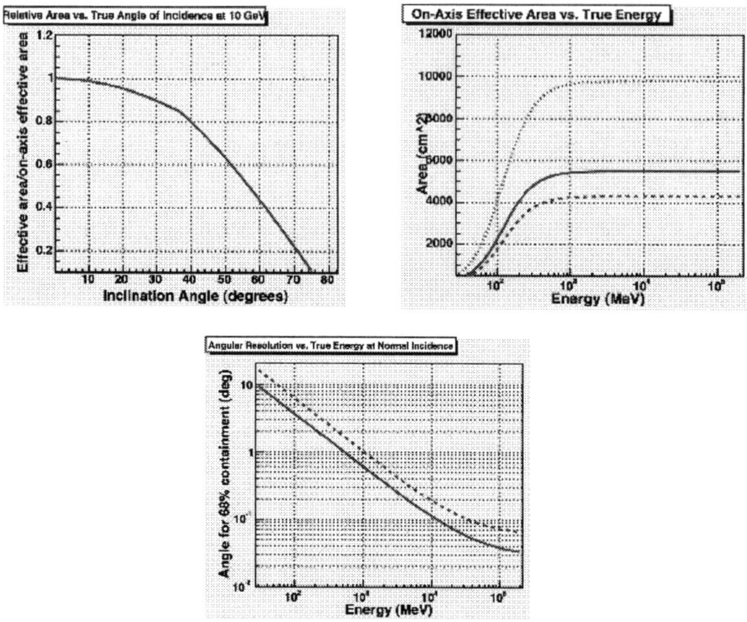

Fig. 2. Exemples of LAT IRFs describing the performance as a function of photon energy and incidence angle.

expected to be better than 1.5^o. GBM spectral response overlaps the LAT one, allowing simultaneous spectral fitting.

4. Scientific purposes

No instrument currently developed explores the entire energy range accessible by the LAT. Moreover, the LAT is a survey instrument, monitoring the entire sky several times ($\sim$ 15 orbits per day with 28.5^o of inclination of the orbital plane) over the course of a single day. While ground-based instruments are now covering part of GLAST energetic range (50 - 300 GeV), they still are suitable for point-source detection but not competitive for survey research. Thanks to these feature GLAST is expected to largely improve the knowledge about a wide range of phenomena:

- Active Galactic Nuclei (AGNs) and Blazars: GLAST will increase the number of known AGN gamma-ray sources from about 100 to thousands;
- Unidentified EGRET Sources: more than 60% of EGRET sources are

unidentified and GLAST should be able to identify them; almost one third of them is expected to be extragalactic (i.e. Blazars) and the rest should be galactic (Radio-quiet Pulsars, Galactic Micro-Quasars, objects associated with the Gould Belt, etc.);

- New Physics: SUSY (in particular the MSSM) theories state the existence of the Lightest Super-Symmetrical Particle: the Neutralino; these theories predict that the annihilation of neutralino and anti-neutralino should give us a γ signal at GLAST energies; recent estimations have indicated GLAST as a very suitable instrument able to detect this signal from the Galactic Center [7] and from the Galactic Halo [8];

- Extragalactic Background Light (EBL): the Diffuse EBL consists of the sum of the starlight emitted by Galaxies through the history of the Universe and a way to study it is to measure the attenuation of γ-ray spectra of distant extragalactic objects by the EBL photons; the sensitivity of GLAST at high energies will permit the measurement of AGN spectra at high energies and hence the study of EBL, distinguishing the intrinsic spectra of AGNs from the effects of attenuation;

- Gamma Ray Bursts (GRBs): LAT and GBM are expected to detect almost 100 GRBs per year, to measure their spectra from keV to GeV energies and to track their afterglows; thanks to its high-energy response and very short deadtime, GLAST is expected to give big improvements in GRB study;

- Pulsars: GLAST will discover many gamma-ray Pulsars, 250 or more, and will provide definitive spectral measurements (in order to distinguish between the two primary models: outer gap and polar cap models);

- Cosmic Rays (CRs) and Interstellar Emission: GLAST, as we will see in next sections, will spatially resolve some SNRs and precisely measure their spectra: so it may prove SNRs as the most important sources of high energy CRs as astrophysical theories predict; spatial and spectral studies on this gamma-ray emission will permit to study separately the distributions of protons and electrons; so GLAST will test CRs production and diffusion theories;

- Solar Flares: EGRET discovered that the Sun is a source of GeV gamma rays; thanks to its Large Effective Area and its small deadtime, GLAST will be able to determine where the acceleration takes place and it should be able to confirm if the protons are accelerated along with electrons.

Another extremely important feature is the coordinate observation by GLAST and ground-based telescopes; this will be of fundamental importance at least for two reasons: the calibration of ground-based telescopes and the multi wavelength observation of variable and steady sources.

5. SNRs simulation

An important strategy for preparing the software for flight is a series of data challenges. Data challenges (DC) are full-sky simulations providing excellent testbeds for science analysis software; teams use data and tools to analyse the simulated science and the "truth" is revealed after a few of weeks to allow crosschecks.

DC2 [9] closed at the beginning of June 2006; it consisted in 55 days of an all-sky survey simulation, adding source variability, extended sources and backgrounds; DC2 includes many Galactic [10] (Diffuse Emission of Milky Way, Pulsars, Plerions, SNRs, XRBs, OB associations, small Molecular Clouds, Dark Matter clumps, the Sun, the Moon and other 3EG sources) and Extragalactic [11] (Galaxies, Galaxy Clusters, AGNs, GRBs, Extragalactic Diffuse γ Background and EBL) sources.

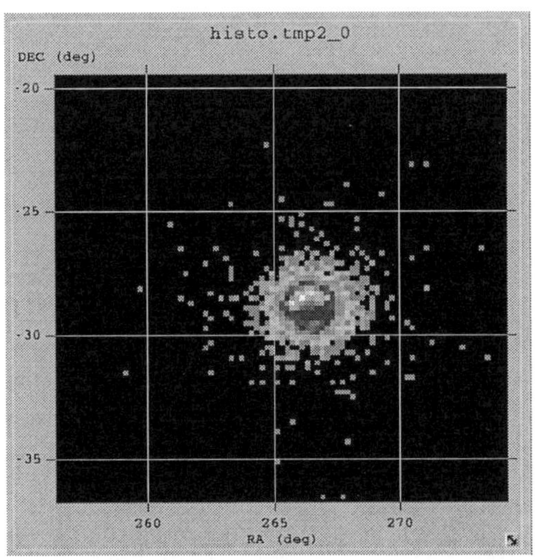

Fig. 3. The interesting case of RXJ0852.0-4622, without backgrounds and other sources, for 10 years of observations.

FIRST RESULTS FROM AMANDA USING THE TWR SYSTEM

ANDREA SILVESTRI[†] for the IceCube Collaboration*

Department of Physics and Astronomy, University of California
Irvine, CA 92697, U.S.A.
[†]*E-mail: silvestr@uci.edu*

The Antarctic Muon And Neutrino Detector Array (AMANDA) has been taking data since 2000 and its data acquisition system was upgraded in January 2003 to read out the complete digitized waveforms from the buried Photomultipliers (PMTs) using Transient Waveform Recorders (TWR). This system currently runs in parallel with the standard AMANDA data acquisition system. Once AMANDA is incorporated into the 1 km^3 detector IceCube, only the TWR system will be kept. We report results from a first atmospheric neutrino analysis on data collected in 2003 with TWR. Good agreement in event rate and angular distribution verify the performance of the TWR system. A search of the northern hemisphere for localized event clusters shows no statistically significant excess, thus a flux limit is calculated, which is in full agreement with previous results based on the standard AMANDA data acquisition system. We also update the status of a search for diffusely distributed neutrinos with ultra high energy (UHE) using data collected by the TWR system.

Keywords: Neutrino Detector, Neutrino Astronomy, Point Sources, Diffuse Sources, Ultra High Energy Neutrinos, AMANDA, TWR, IceCube

Introduction

Neutrinos are the only high energy particles able to propagate undeflected and unattenuated from the furthest reaches of the Universe. Extragalactic UHE γ-ray astronomy falters for energies greater than a few tens of TeV due to interactions with infrared and Cosmic Microwave Background photons. The information carried by the neutrino messengers from distant, unexplored regions of the universe may help to unravel longstanding mysteries associated with the origin of the highest energy cosmic rays. Several models[1, 2, 3, 4, 5, 6] predict high energy charged particle and neutrino emission; the diffuse ν-flux predictions by these models are constrained by

*http://www.icecube.wisc.edu

the observed cosmic ray fluxes[7]. AMANDA[8], the first neutrino telescope constructed in transparent ice, is deployed between 1500 m and 2000 m beneath the surface at the geographic South Pole. It is designed to search for neutrinos that originate in the most violent phenomena in the observable universe. Galactic objects like Supernova Remnants (SNR) and extragalactic objects such as Active Galactic Nuclei (AGN) are expected to be the powerful engines accelerating protons and nuclei to the highest energies, which eventually interact to generate neutrinos. We have searched for point sources in the northern sky and for a diffuse flux of UHE ($E_\nu > 10^{15}$ eV) neutrinos of cosmic origin. We have performed the analysis using data collected in 2003 with a new data acquisition system based on Transient Waveform Recorders (`TWR-DAQ`), which we compared to data collected by the standard AMANDA readout system (μ-`DAQ`).

Fig. 1. The upgraded AMANDA data acquisition electronics with Transient Waveform Recorders, here displayed in a system of 2 racks with 72 TWR modules.

1. The TWR system

The data acquisition electronics of the AMANDA-II detector (Fig. 1) was upgraded in 2003 to read out the complete digitized waveform of the photomultiplier tubes (PMTs) using Transient Waveform Recorders (TWR)[9]. The transition was made to run the μ-`DAQ`[10] and `TWR-DAQ` in parallel to verify that the performance of the latter is as good or better. To compare the two systems, the data from 2003 has been analyzed with data sets from both readout systems. The performance of the TWR-DAQ is verified by comparing the results for the absolute rate of atmospheric neutrinos and the $cos(\theta)$ distribution with the results from the standard μ-DAQ[11]. Extending the analysis tools to include TWR data required several new developments: (1) The TWR-DAQ measures the integrated charge Q from the full waveform of the PMT pulses, in contrast, the μ-DAQ only measures the maximum amplitude of the PMT pulse in a 2 μs window. (2) Various time offsets were taken into account for the TWR-system, none of which impacted the μ-DAQ. A time resolution of few nanoseconds was extracted. (3) The size of the TWR-DAQ data set (20TB for 2003) by far exceeds that of the μ-DAQ ($\sim$1TB), requiring an improved data handling tools.

2. Analysis

The data information of the two systems has been merged according to GPS time and the fraction of overlapping PMTs participating in the event in both systems. For the point source analysis we restricted the capabilities of the TWR-DAQ system to mimic the features of the μ-DAQ system as closely as possible. The timing and amplitude information extracted from waveforms has been used as input parameters to perform PMT-pulse cleaning, TOT (Time-Over-Threshold) and crosstalk[a] cleaning. A Gaussian fit of the

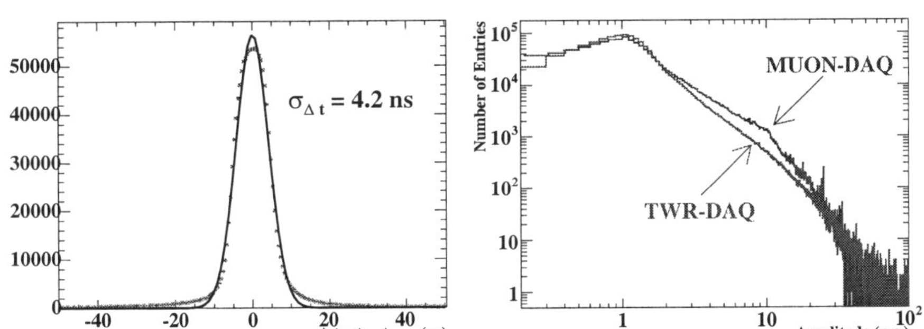

Fig. 2. (Left) Time difference between the PMT pulses recorded by the two acquisition systems $\Delta t = t_\mu - t_{TWR}$. (Right) Calibrated amplitude normalized to 1 photo-electron (p.e.) value for μ-DAQ and TWR-DAQ data. See text for details.

[a]Any phenomenon by which a signal transmitted on one channel of a transmission system creates an undesired effect in another channel.

distribution for $\Delta t = t_\mu - t_{TWR}$ yields $\sigma_{\Delta t} = 4.2$ ns (Fig. 2 (left)), which is dominated by the systematic error of the time jitter between independent flash ADC clocks of the TWR system. The TWR-DAQ timing calculations are relative to the values measured by the μ-DAQ. The timing of the TWR-DAQ system includes two sources of jitter, the digitization window of 10 ns and the time fluctuation of the flash ADC clocks within the same 10 ns interval. This accounts for the $\sigma_{\Delta t} \sim 4$ ns[b]. Amplitudes are also calibrated by extracting the number of photo-electrons (Npe) detected from the peak ADC of the μ-DAQ and charge $\mathcal{Q}$ of the TWR system and normalized to 1-pe amplitude. By integrating the charge from pulses in the waveform, the dynamic range of the TWR-DAQ extends to Npe of ~ 100, a factor of three higher than the μ-DAQ. Fig. 2 (right) shows the reconstructed amplitude of the TWR-DAQ compared to the μ-DAQ, which indicates a stable power law distribution extending up to 100 Npe, while the μ-DAQ system shows a "knee" around 10 Npe. The knee is due to the amplitude saturation of channels read out by optical fibers, which comprise ~40% of the AMANDA channels.[c] After cleaning, the muon track is reconstructed from the pulse times of the PMTs. Details on the reconstruction techniques can be found in[12]. Table 1

Table 1. Passing rates for increasing level of data selection criteria for the TWR-DAQ and μ-DAQ data analysis.

Selection	TWR-DAQ	μ-DAQ
Level-0 (Raw)	1.86×10^9	1.86×10^9
Level-1	1.25×10^8	1.25×10^8
Level-2	2.56×10^6	1.99×10^6
Level-3 (Final)	1112	1026

summarizes the passing rates from the raw data level to the final sample of atmospheric neutrinos. The event selection criteria used in this analysis follows the method described in[13]. Fig. 3 shows that the angular mismatch for the final neutrino sample $\sigma_{\Delta\theta}$ is 0.5° where $\Delta\theta = \theta_\mu - \theta_{TWR}$. This value is expected from studies of the precision of the global minimizer in the reconstruction program, by reconstructing the same event sample twice with a 32-iteration algorithm. The azimuth angular mismatch $\sigma_{\Delta\phi}$ is 0.7°, which

[b]For year 2005 the phases of the TWR modules were synchronized to avoid the time jitter between independent flash ADC clocks.

[c]High voltage values were lowered in January 2005 to increase linear dynamic range of optical channels.

is consistent with a σ of 0.5° for the $\Delta\phi \times \sin(\theta)$ distribution to account for the zenith dependence on the azimuth angle.

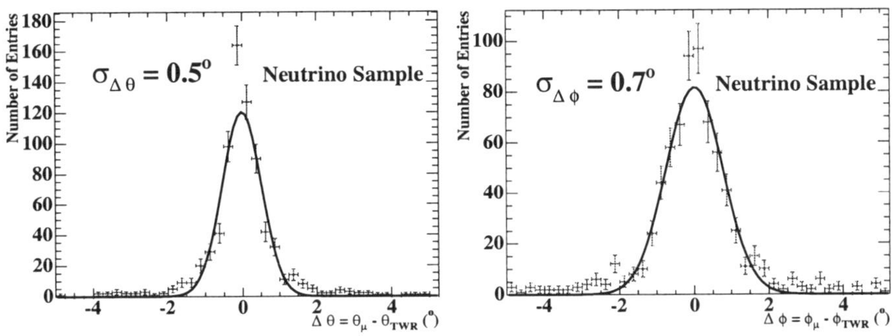

Fig. 3. (Left) The zenith $\Delta\theta = \theta_\mu - \theta_{TWR}$ difference distribution between the μ-DAQ and the TWR-DAQ systems, (right) the azimuthal $\Delta\phi = \phi_\mu - \phi_{TWR}$ difference distribution.

3. Search for point sources of neutrinos

From the analysis based on TWR-DAQ data, 1112 neutrinos are observed compared to 1026 neutrinos from the μ-DAQ data analysis. The small differences in the event rate are compatible with the small differences in analysis procedures described in Section 2. Fig. 4 shows the cos(θ) distribution of the

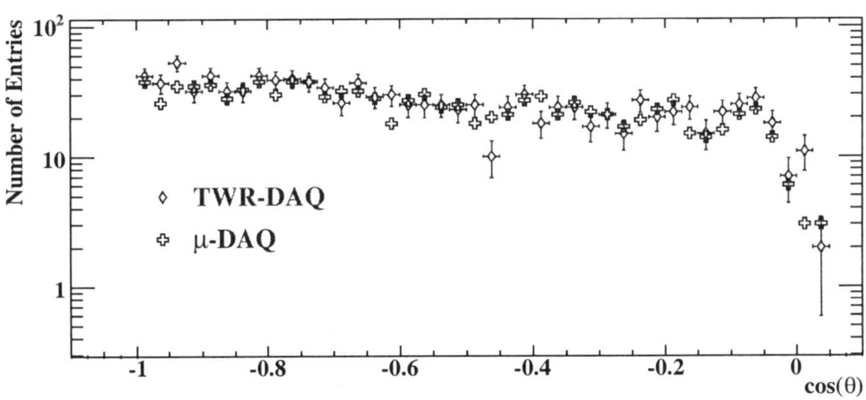

Fig. 4. Distribution of cos(θ) after final selection representing the atmospheric neutrino sample observed from the TWR-DAQ and the μ-DAQ analyses.

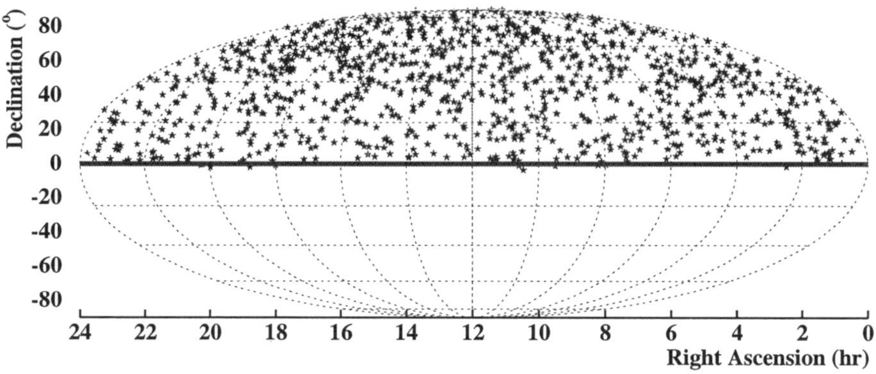

Fig. 5. Sky plot displayed in equatorial coordinates, Right Ascension (RA) and Declination (DEC), of the final sample of 1112 atmospheric neutrinos observed from the TWR-DAQ analysis.

atmospheric neutrino sample extracted from the TWR-DAQ and μ-DAQ data analysis. Satisfactory agreement can be seen for the $\cos(\theta)$ distribution of the atmospheric neutrinos samples obtained by the different analyses. Fig. 5 displays the sky map plotted in equatorial coordinates of the 1112 atmospheric neutrino candidates observed from the TWR-DAQ analysis. All events are distributed in the norther hemisphere with very few events reconstructed at the horizon. In order to distinguish if the observed events follow a random distribution as expected by atmospheric neutrinos, or are an indication of localized event cluster as expected by a source of neutrinos

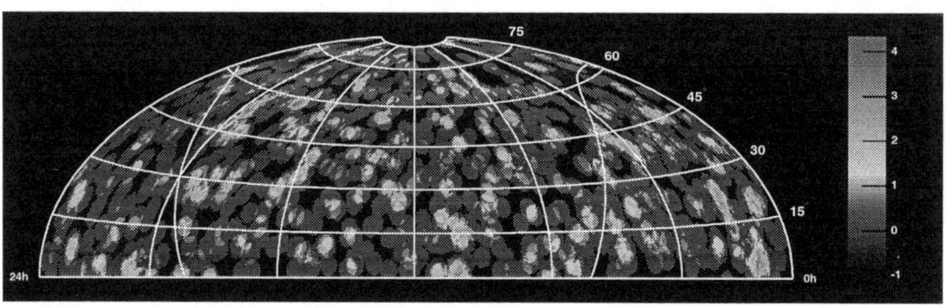

Fig. 6. Sky map plotted in coordinates of right ascension and declination of the 1112 atmospheric neutrino candidates from the TWR-DAQ analysis. The scale on the right reflects excess or deficit in terms of the positive or negative deviation with respect to the mean background events.

of extraterrestrial origin, an analysis is required to estimate the statistical significance of the observed events. A full search of the northern sky was performed to look for any localized event cluster. The full scan is extended to 85° in declination, since the limited statistics in the polar bin prevents an accurate estimate of the background. Fig. 6 shows the sky map of the calculated significance for all observed 1112 events in terms of the a positive or negative deviation with respect to the mean background events. All observed regions with the highest significance are compatible with the background hypothesis. The highest significance observed shows a positive deviation of 4.3 σ, which corresponds to a probability of 23% for a search performed on events with randomized right ascension.

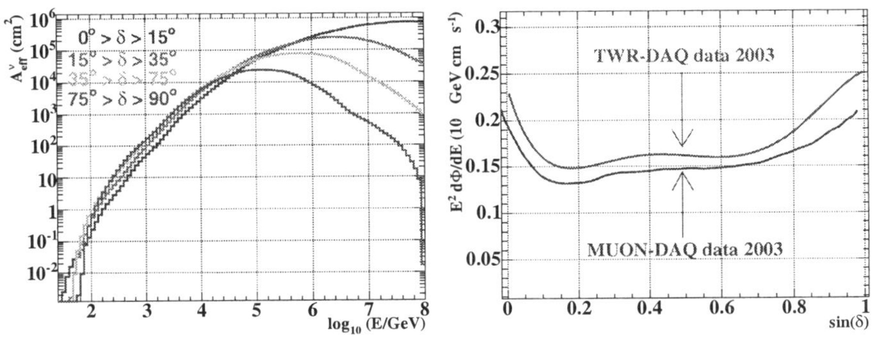

Fig. 7. (Left) Muon neutrino effective area as a function of E_{ν_μ} for point source searches displayed for different declinations δ. (Right) Average upper limit as a function of $\sin(\delta)$ on ν_μ for a E^{-2} neutrino flux spectrum from the TWR-DAQ and the μ-DAQ analyses.

Numerous studies have been performed to implement the best possible description of the TWR system in the Monte Carlo simulation. In particular, a correct description of the waveform hardware response is now available and proper TDC time windows as well as amplitude calculations have been now implemented. Therefore a calculation of the neutrino effective area for point source search using TWR data is now feasible. Fig. 7 (left) shows the effective muon neutrino area as a function of E_{ν_μ} computed at the final selection of this analysis for four separated ranges of the declination. The A^{ν}_{eff} extends over six order of magnitudes as ν-energy increases, while A^{ν}_{eff} decreases as declination increases due to the neutrino absorption in the Earth. The average upper limit on ν_μ as a function of declination is shown in the right panel of Fig. 7. The average neutrino flux upper limit is determined from the ratio of the average Feldman and Cousins[14] upper limit $< \mu^{90} >$ computed according to the expected mean background and observed events, and the number of the expected signal events for a neutrino flux $E_\nu^2 d\Phi_\nu/dE_\nu =$ GeV $s^{-1}cm^{-2}$ from a point source at declination δ. Together with the average upper limit extracted from the TWR-DAQ data, the limit from the μ-DAQ data of year 2003[13] is also represented for comparison. These results are comparable, however the TWR-DAQ limit is $\sim 10\%$ worse than the μ-DAQ limit, because conservative event selection criteria have been applied that do not use the full waveform information in an optimal way. The preliminary average upper limit from the TWR-DAQ analysis on the muon neutrino flux with spectrum $d\Phi_\nu/dE_\nu \propto E_\nu^{-2}$ is $E_\nu^2 d\Phi_\nu/dE_\nu \leq 1.8 \times 10^{-7}$ GeV $s^{-1}cm^{-2}$ at 90% confidence level, in the energy range 1.26 TeV $< E_\nu <$ 1.6 PeV. This upper limit does not include systematic uncertainties. AMANDA has submitted a publication on a 5-year search for point sources (2000-2004) based on the μ-DAQ data[13], which improves the limit by ~67% compared to the results from this analysis.

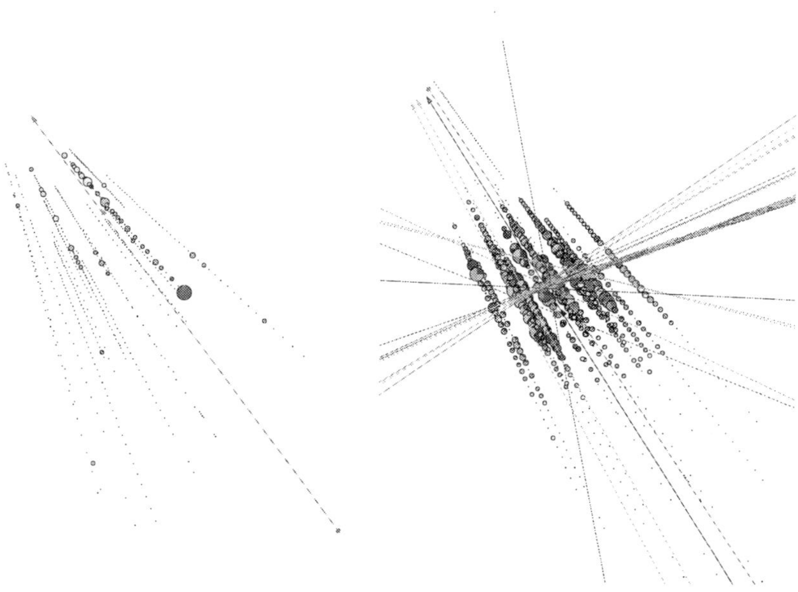

Fig. 8. (Left) one of the 1112 atmospheric neutrino event detected in the experimental array. (Right) a high-energy MC simulated event. The size of the circles represent energy deposited in the detector, the colors represent the time profile of photons propagating in the ice which have scattered to the PMTs.

4. Search for UHE neutrinos

The search for UHE neutrinos differs from the point source mainly in two aspects. First, the search for point sources is restricted to upward going events of energy within the TeV scale, the UHE analysis extends the search beyond PeV energies. Second, while point sources are localized in a small region of the sky, UHE diffuse neutrinos are expected to come from the entire visible sky (2π sr), with a higher probability to originate from the horizon. Fig. 8 shows one of the 1112 upward going neutrino events and an almost horizontal UHE simulated neutrino event with energy E $\sim 3 \times 10^{15}$ eV. Simulated events with energy above one PeV deposit a substantial amount of light which is recorded by almost all PMTs. In order to better separate background events from signal events, we developed new variables which exploit the information of the full waveforms. The capability of improved

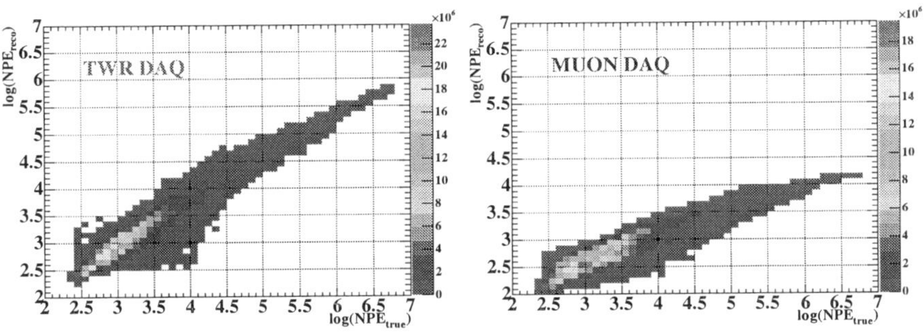

Fig. 9. Contour plots displaying the capability of correctly reconstructing the total number of photon-electron injected into the detector (left) for the TWR-DAQ and (right) for the μ-DAQ system.

photo-electron reconstruction of the TWR-DAQ system can be seen in Fig. 9, which displays a contour plot in $\log_{10}$-scale of the sum of the reconstructed number of PE, NPE_{reco} versus the NPE_{true} generated by the signal MC simulation. While the μ-DAQ system quickly saturates on the order of few thousands of PE's, the TWR-DAQ, by using the after-pulse information, can see the millions of PE's which are typical for UHE events triggered by single muons. Fig. 10 summarizes the results of several AMANDA searches for diffuse neutrinos at different energy ranges using μ-DAQ data. The experimental limits assume a 1:1:1 ratio of neutrino flavors at the Earth due to oscillation. The dotted and dashed lines represent a sample of model predictions[1, 2, 3, 4, 6], adjusted for oscillation if necessary. The horizontal solid lines represent limits on the integrated neutrino flux from the diffuse analyses with AMANDA-B10[15], with AMANDA-II[16], from the cascade analysis[17], and from the UHE analysis[18, 19], with the NT200 neutrino telescope[20], and at the highest energies from ANITA-lite[21] and RICE[22]. These are the most stringent flux limits at 90% C.L. for an E^{-2} spectrum to date. The AMANDA limits have been determined by analysing data collected with the μ-DAQ system. However, as it has been shown above, by performing a new analysis approach with TWR-DAQ data, it is possible to develop new techniques and to improve the current experimental limits for the UHE neutrino search.

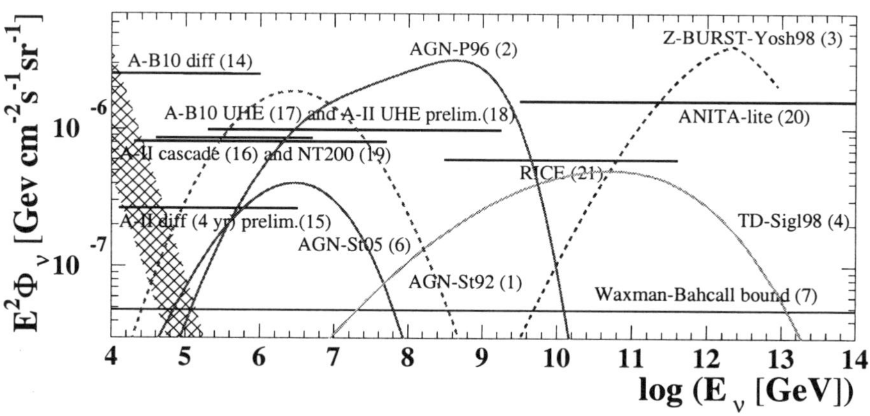

Fig. 10. Experimental limits on the integrated neutrino flux and detector sensitivity for a neutrino flux of E^{-2} spectrum. The curves represents predictions from theoretical models, the solid lines show current AMANDA (all-flavor limits), NT200, ANITA and RICE experimental limits.

5. Conclusion

The point source search provides the first detailed evaluation of the performance of the TWR-DAQ system. This analysis demonstrated that the TWR-DAQ produces similar event rates and angular distribution as the data collected by the standard μ-DAQ system. No statistically significant excess has been observed after performing a full search of the northern hemisphere of the sky for localized event clusters, therefore a flux limit based on TWR-DAQ data is calculated. The AMANDA TWR readout is now being incorporated into the IceCube data acquisition system. By exploiting the information of the full waveforms from the `TWR-DAQ` system, it is possible to develop new analysis techniques for an improved search for diffuse UHE neutrinos. Currently the AMANDA experiment has placed the most stringent neutrino flux limits to date, which can be further improved by analyses performed with `TWR-DAQ` data.

Acknowledgments

The author acknowledges support from the U.S. National Science Foundation (NSF) Physics Division, the NSF-supported TeraGrid systems at the San Diego Supercomputer Center (SDSC) and the National Center for Supercomputing Applications (NCSA), the Phi Beta Kappa Alumni in Southern California for providing travel grants, the Astrophysics Associates, Inc., Italian Ministry of Education, European Physical Society, Ettore Majorana Foundation, and the Electron Tubes Ltd. for providing the full scholarship at Erice.

References

1. F. Stecker et al., *Phys. Rev. Lett.* **66**, 2697 (1992).
2. R. J. Protheroe, *astro-ph/9607165* (1996).
3. S. Yoshida et al., *Phys. Rev. Lett.* **81**, 5505-5508 (1998).
4. G. Sigl, *Phys. Rev. D* **59**, 043504 (1999).
5. R. Engel et al., *Phys. Rev. D* **64**, 093010 (2001).
6. F. W. Stecker, *Phys. Rev. D* **72**, 107301 (2005).
7. E. Waxman and J. Bahcall, *Phys. Rev. D* **59**, 023002 (1999).
8. A. Silvestri et al., in *Neutrinos and Explosive Events in the Universe*, Springer, New-York, **209**, 275-285 (2005).
9. W. Wagner et al., in *28th ICRC*, **2**, 1365-1368 (2003).
10. R. Wischnewski et al., in *27th ICRC*, **2**, 1105-1108 (2001).
11. A. Silvestri et al., in *29th ICRC*, **5**, 431-434 (2005).
12. J. Ahrens et al., *Nucl. Instrum. Meth.* **A524**, 169 (2004).
13. A. Achterberg et al., *astro-ph/0611063* (2006).
14. G. J. Feldman and R. D. Cousins, *Phys. Rev. D* **57**, 3873 (1998).
15. J. Ahrens et al., *Phys. Rev. Lett.* **90**, 251101 (2003).
16. G. Hill et al., in Proc. of the XXII International Conference on Neutrino Physics and Astrophysics (2006).
17. M. Ackermann et al., *Astropart. Phys.* **22**, 127 (2005).
18. M. Ackermann et al., *Astropart. Phys.* **22**, 339 (2005).
19. L. Gerhardt et al., in Proc. of SUSY06 (2006).
20. V. Aynutdinov et al., *Astropart. Phys.* **25**, 140 (2006).
21. S. Barwick et al., *Phys. Rev. Lett.* **96**, 171101 (2006).
22. I. Kravchenko et al., *Phys. Rev. D* **73**, 082002 (2006).

NEMO: A PROJECT FOR A KM³ UNDERWATER DETECTOR FOR ASTROPHYSICAL NEUTRINOS IN THE MEDITERRANEAN SEA

Isabella Amore* for the NEMO Collaboration

Department of Physics and Astronomy - University of Catania
Laboratori Nazionali del Sud - INFN
Via S. Sofia 62 I-95123 Catania, Italy
**E-mail: amore@lns.infn.it*

The status of the project is described: the activity on long term characterization of water optical and oceanographic parameters at the Capo Passero site candidate for the Mediterranean km^3 neutrino telescope; the feasibility study; the physics performances and underwater technology for the km^3; the activity on NEMO Phase 1, a technological demonstrator that is going to be deployed at 2000 m depth 25 km offshore Catania; the realisation of an underwater infrastructure at 3500 m depth at the candidate site (NEMO Phase 2).

Keywords: high energy neutrino astronomy, underwater telescopes

1. Introduction

The construction of a km^3 underwater detector for high energy astrophysical neutrinos is one of the main goals of astroparticle physics. High energy neutrinos are very promising probes for the investigation of the far universe and of the acceleration processes occurring in galactic and extragalactic sources. Differently from charged particles and gamma rays with $E_\gamma >$TeV, neutrinos can reach the Earth from far away cosmic accelerators, travelling in straight line, thus carrying direct information on the source. Theoretical models indicate that a detection area of $\simeq$1 km^2 is required for the measurement of HE cosmic ν fluxes. The underwater/ice Cherenkov technique is considered the most promising experimental approach to build high energy neutrino detectors. The first generation of underwater/ice neutrino telescopes, BAIKAL [1] and AMANDA [2], despite their limited size have already set first constraints on astrophysical models of TeV neutrino sources. The successful experience of AMANDA opened the way to Ice-Cube [3], a km^3-size neutrino telescope which is now under construction at the South Pole. ANTARES [4] is now taking data with 2 lines and others will soon be deployed. The scientific community is supporting the construction of another km^3 neutrino telescope in the Northern Hemisphere, to allow contemporary observation of the full sky. The Mediterranean Sea offers optimal conditions to locate the telescope and a Mediterranean km^3 telescope should be able to observe neutrinos from the Galactic Centre region, not seen by IceCube. For this reason the EU founded Km3Net [5] which is expected to provide a design study for the Mediterranean km^3 telescope. The NEMO collaboration is carrying out, since 1998, an *R&D* programme towards this purpose. The activities of the group have been focused on three items: the search and long term monitoring of an optimal site for the installation of the km^3 telescope; the development of a technical and scientific feasibility study of the detector; the realisation of a technological demonstrator (the NEMO Phase 1 project), which is scheduled to be deployed and operated by the end of this year.

2. Site selection and characterization

The km^3 detector needs first a complete knowledge of the site physical and oceanographical characteristics over a long time period. The NEMO collaboration performed, since 1998, a long term research program to find and characterise an optimal deep sea site. After 30 sea campaigns a site located in the Ionian Sea (36° 19' N, 16° 05' E), close to Capo Passero (South-East of Sicily), was identified as an optimal candidate. The site is a wide abyssal plateau with an average depth of about 3500 m, located at less than 80 km from the shore and about 50 km from the shelf break (see Figure 1). Water transparency was measured in situ using a set-up based on a transmissometer that allowed to measure light absorption and attenuation at nine different wavelengths ranging from 412 to 715 nm. The values of the light absorption length measured at depths of interest for the detector installation (more than 2500 m) are close to the one of optically pure sea water (about 70 m at $\lambda = 440$ nm). Seasonal variations are compatible with the instrument experimental error [6]. Another characteristic of deep sea water that can affect the detector performance is the optical background. This comes from two natural causes: the decay of the ^{40}K, that is present in seawater, and the bioluminescence light produced by biological organisms like bacteria or fish. In Capo Passero an average rate of about 30 kHz of optical noise (measured with 10" PMTs at 0.3 single photoelectron threshold) was measured at a depth of 3000 m in several sea campaigns. This value is compatible with what expected from pure ^{40}K background, with rare high rate spikes due to bioluminescence, in agreement with the measured vertical distribution of bioluminescent organisms (the measurement indicate a low concentration of luminescent organisms at depths greater than 2500 m). The programme of site characterization also includes long term measurements of water temperature and salinity, deep sea currents, sedimentation rate and bio-fouling. All data confirm the optimal characteristics of the site [7].

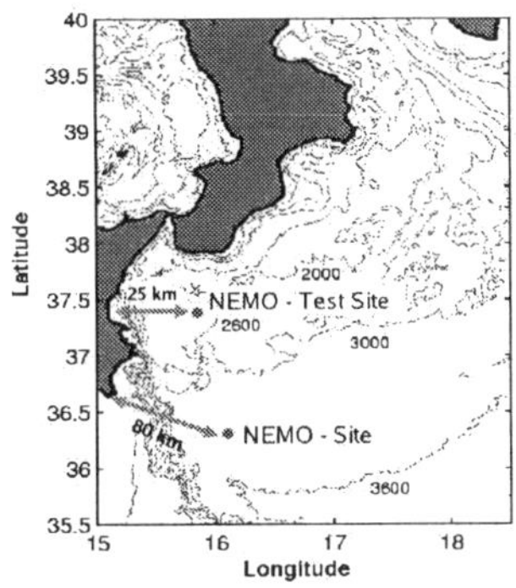

Fig. 1. Bathymetric chart of Eastern Sicily. The locations of the Capo Passero site and of the NEMO Phase 1 *Test Site* are shown. The seabed depth is about 3500 m for the *Capo Passero* site and 2000 m for the Test Site.

3. Feasibility study for the km^3 telescope

The design of an underwater km^3 neutrino telescope represents a challenging task; one has to optimise the physics discovery potentials, technical feasibility and budget. A km^3 neutrino telescope is defined as an array of about 5000 optical modules (OM) hosted on underwater structures. NEMO proposes innovative structures: the NEMO *towers* designed to deploy, during a single operation, a large number of OMs. Each floor of the tower hosts four optical modules with 10" PMTs. Floors are arranged in a 3-dimensional structure in order to locally allow event trigger and track reconstruction. The proposed detector geometry consists of a squared array of 9×9 towers made of 16 floors each and with 5184 OMs. The distance between the towers is 140 m. The detector performances were evaluated by means of numerical simulations, using the software developed by the ANTARES collaboration [8] and adapted to a km^3 scale detectors [9]. Using the above geometry and the site parameters measured in Capo Passero, Monte Carlo simulations show that the detector can reach an effective area >1 km^2 at muon energies of about 10 TeV (see Figure 2 Right), with an angular resolution of the order of few tenths of a degree. Simulations also show that the expected sensitivity (see Figure 2 Left) to a point like astrophysical source at declination =-60° is about 1.2 10^{-9} GeV cm^{-2} s^{-1} (the used spectrum is $E_{\nu_\mu}^{-2}$), obtained for a search bin of 0.3 degrees [10]. The simulations indicate that such a detector may reach a better sensitivity and smaller search bin than IceCube.

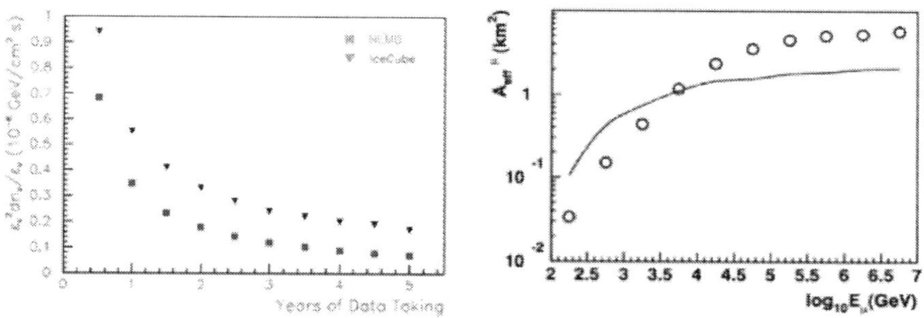

Fig. 2. Right: Muon effective areas as a function of the muon energy for square lattices of NEMO towers 140m spaced (continuous line) and 300m spaced (open circles). Left: Sensitivity to a neutrino spectrum with $\alpha = 2$, coming from a $\delta = -60°$ declination point-like source and comparison with the IceCube detector [3].

4. NEMO Phase 1 and Phase 2

In order to validate the characteristics of the mechanical structure, of the data transmission and of the power distribution system, we are developing a technological demonstrator, called NEMO Phase 1, that includes prototypes of the critical elements of the proposed km^3 detector. The project is under realisation at the Test Site of the LNS at the depth of 2000 m close to Catania; a 25 km electro optical cable allows the connection of the deep sea instrumentation to a shore station, located inside the port of Catania. The shore laboratory hosts the power feeding system, the instrumentation control system, the land station of the data transmission system and the data acquisition. NEMO Phase 1 will deploy and operate a Junction Box (JB) and a "mini-tower" (see Figure 3). The Junction Box provides connection between the main electro-optical cable and the instruments in the tower. It is designed to host and protect, from the effects of corrosion and pressure, the opto-electronic boards dedicated to the distribution and con-

Fig. 3. The Junction Box (Left) and the mini-tower (Right) under assembling.

trol of the power supply and digitized signals. A jumper cable will connect the JB to the mini-tower made of 4 floors. Each floor has a rigid aluminium structure, 15 m long, that hosts 4 OMs, environmental and control sensors, the front-end electronics. The vertical distance between floors is 40 m. Floors are interlinked by a system of cables; the tower line is anchored on the seabed and is kept vertical by a buoyancy on the top. After deployment the structure is unfurled: each floor will be rotated by 90°, with respect to the up and down adjacent ones, around the vertical axis of the tower. One of the advantages of this structure is the fact that it can be compacted, by piling each storey upon the other, to allow transport and deployment. Inside each floor are installed a Power Floor Control Module (PFCM) and a Floor Control Module (FCM). The latter is the core of the system since it hosts all the floor electronics for data transmission. The FCM is connected to the OMs by means of four electro-optical cables, and to auxiliary instrumentation (oceanographic probes, hydrophones for the acoustic positioning system) via electrical cables. Each optical module is composed by a 10" photo-multiplier enclosed in a 17" pressure resistant sphere of thick glass. In spite of its large photocathode, the used PMT has a time resolution of $\sim$3 ns FWHM for single photoelectron pulses and a charge resolution of $\sim$35%. The base card circuit for the high voltage distribution has a power consumption of less than 150 mW. A front-end electronics board is also placed inside the OM. Sampling at 200 MHz is accomplished by two 100 MHz staggered Flash ADCs, whose outputs go to an FPGA and transmitted to the FCM at 20 Mbit/s rate. The Phase 1 detector subsystems have undergone extensive qualification and pressure tests. Phase 1 is now integrated at the Test Site shore laboratory; final tests of the complete system are running, the deployment is scheduled for the end of this year. Although the Phase 1 project will provide an important test of the technologies proposed for the realisation of the detector, these must be finally validated at the depths needed for the km^3 detector. For these motivations the collaboration is planning the realisation of an infrastructure at the Capo Passero site (NEMO Phase 2). It will consist of a shore laboratory (located inside the harbour area of Capo Passero), 100 km underwater electro-optical cable (linking the 3500 m deep sea site to the shore) and the underwater infrastructures needed to connect prototypes of the km^3 detector.

5. Conclusions

The design of the Mediterranean km^3 telescope for high energy astrophysical neutrinos is a challenging task; several collaborations in Europe are working on the realisation of demonstrators. More efforts are needed to develop a project for the km^3 detector, for this reason EU funded Km3Net, which is expected to provide a design study for the Mediterranean km^3 telescope within 2009. The activities of the NEMO collaboration are contributing to this goal: an extensive study on site properties has demonstrated that Capo Passero has optimal characteristics for the telescope installation; a technical feasibility analysis for the km^3 detector has shown that a detector with effective area for TeV muons $\sim$1 km^2 is realisable at an affordable cost; a technological demonstrator is being installed at the underwater Test Site of the LNS in Catania.

References

1. Wischnewski R., BAIKAL Collaboration 2005, *Proc. of 29th ICRC* (Pune, India)
2. Ackermann M. et al., 2005 *Phys. ReV.* **D077102**
3. Ahrens J. et al., *Astrop. Phys.* **20**(2004)507.
4. Aguilar J. A. et al., *Astrop. Phys.* **26**(2006)314, astro-ph/0606229.
5. `http://www.km3net.org`
6. Riccobene G. et al. 2006 (astro-ph/0603701), accepted for pubblication in *Astrop. Phys.*
7. The NEMO Collaboration 2003, *Study and characterization of a deep sea site for a km^3 underwater neutrino telescope,* `http://nemoweb.lns.infn.it/sites/sitereport`
8. Becherini Y., ANTARES Collaboration, *Proc. VLVNT2 Workshop (Catania)*
9. Sapienza P., NEMO Collaboration 2003, *Proc. VLVNT Workshop (Amsterdam)* `http://www.vlvnt.nl/proceedings`
10. Distefano C., NEMO Collaboration 2006, *Proc. BCN06 (Barcelona)* `http://www.am.ub.es/bcn06/`

RESULTS FROM THE ANITA EXPERIMENT

ANDREA SILVESTRI† for the ANITA Collaboration*

Department of Physics and Astronomy, University of California
Irvine, CA 92697, U.S.A.
† E-mail: silvestr@uci.edu

*The ANITA Collaboration:

S.W.Barwick[3], J.J.Beatty[7], D.Z.Besson[6], W.R.Binns[8], B.Cai[10], J.M.Clem[1], A.Connolly[4], D.F.Cowen[2], P.F.Dowkontt[8], M.A.DuVernois[10], P.A.Evenson[1], D.Goldstein[3], P.W.Gorham[5], C.L.Hebert[5], M.H.Israel[8], J.G.Learned[5], K.M.Liewer[9], J.T.Link[5], S.Matsuno[5], P.Miocinovic[5], J.Nam[3], C.J.Naudet[9], R.Nichol[7], K.Palladino[7], M.Rosen[5], D.Saltzberg[4], D.Seckel[1], A.Silvestri[3], B.T.Stokes[5], G.S.Varner[5], F.Wu[3].

[1] *Bartol Research Institute, University of Delaware, Newark, Delaware,* [2] *Dept. of Astronomy and Astrophysics, Penn. State University, University Park, Pennsylvania,* [3] *Dept. of Physics and Astronomy, University of California, Irvine, California,* [4] *Dept. of Physics and Astronomy, University of California, Los Angeles, California,* [5] *Dept. of Physics and Astronomy, University of Hawaii, Manoa, Hawaii,* [6] *Dept. of Physics and Astronomy, University of Kansas, Lawrence, Kansas,* [7] *Dept. of Physics, Ohio State University, Columbus, Ohio,* [8] *Dept. of Physics, Washington University in St. Louis, Missouri,* [9] *Jet Propulsion Laboratory, Pasadena, California,* [10] *School of Physics and Astronomy, University of Minnesota, Minneapolis, Minnesota*

The ANtarctic Impulse Transient Antenna (ANITA) is the first long-duration balloon experiment designed to search and measure the flux of Greisen-Zapsepin-Kuzmin (GZK) neutrinos. We present new limits on neutrinos fluxes of astronomical origin from data collected with the successful launch of a 2-antenna prototype instrument, called ANITA-lite, that circled the Antarctic continent for 18.4 days in January 2004. We performed a search for Ultra-High-Energy (UHE) neutrinos with energies above 3×10^{18} eV. No excess events above the background expectation were observed and a neutrino flux following E^{-2} spectrum for all neutrino flavors, is limited to $E_{\nu}^{-2}F < 1.6\times 10^{-6}$ GeV cm^{-2} s^{-1} sr^{-1} for $10^{18.5}$ eV $< E_{\nu} < 10^{23.5}$ eV at 90% confidence level. The launch of ANITA is scheduled for December 2006. Looking beyond ANITA, we describe a new idea, called ARIANNA (Antarctic Ross Iceshelf ANtenna Neutrino Array), to increase the sensitivity for GZK neutrinos by one order of magnitude better than ANITA.

Keywords: High Energy Neutrinos, Neutrino Astronomy, Antarctic Ice Attenuation, GZK Neutrinos, Radio Detection, ANITA, ARIANNA

Introduction

ANITA is designed to search for neutrinos that are originated in the most violent phenomena in the universe. Its primary mission is to discover GZK neutrinos generated by interactions between protons with energy above 5×10^{19} eV and the cosmic microwave background radiation. Flying at an altitude between 35-40 km above the Antarctic continent, the ANITA balloon-borne telescope will be sensitive to a target volume of $\sim 10^6$ km^3 of radio-transparent ice. The aperture of ANITA exceeds ~ 100 km^3·sr at $E_\nu = 3\times 10^{18}$ eV, averaged over neutrino flavor and assuming equal fluxes of all flavors. The aperture continues to grow rapidly as E_ν increases, reaching the order of 10^5 km^3·sr at $E_\nu = 10^{21}$ eV. At the energies of relevance to ANITA, the Earth strongly attenuates the neutrino flux, so ANITA is primarily sensitive to horizontal neutrinos (Fig. 1 left). Neutrinos interacting within the ice generate compact particle showers, which emit coherent radio signals that are detectable in the radio-quiet environment of the Antarctic continent. The neutrino radio signal can also be used to measure the neutrino cross-section at the highest energy. ARIANNA is a new idea born to explore the GZK neutrino fluxes with a sensitivity one order of magnitude greater than ANITA, and also to fill the energy gap for the neutrino search between $10^{17} - 10^{18}$ eV.

1. Science motivation

High energy neutrinos carry unique information from objects in the universe, and complement the information provided by UHE γ-ray astronomy. For example, neutrinos, unlike photons, can propagate throughout the universe unattenuated, but photon astronomy between 10-100 TeV is limited to distances less than a few hundred Mpc due to interactions with infrared background photons, and to even shorter distances at PeV energies due to interactions with the cosmic microwave background. Neutrinos may be the only method to shed light on acceleration processes associated with sources of the highest energy cosmic rays, which extend more than six orders of magnitude above energy of 100 TeV. Thus, the direct detection of high energy neutrinos [1, 2, 3] will complement the investigation of the GZK cutoff [4, 5], one of the most controversial issues in cosmic ray physics. The ANITA science goals extend beyond the Standard Model if GZK neutrinos produce highly unstable micro black holes (BH) when they interact with ice [6, 7]. The decay of these highly unstable micro BH via Hawking radiation would generate energetic hadronic showers that ANITA would detect. The signature of such events is the observation of an enhanced detection rate that is strongly energy dependent. This observation would provide evidence of new phenomena, such as the existence of extra dimensions.

2. ANITA

The concept of detecting high energy particles through coherent radio emission was first postulated by Askaryan [8] and has recently been confirmed in accelerator experiments [9]. Particle cascades induced by neutrinos in Antarctic ice are very compact, no more than a few centimeters in lateral extent [10]. The resulting radio emission is coherent Cherenkov radiation which is characterized by a conical emission geometry, broadband frequency content, and linear polarization. ANITA [11] will observe the Antarctic ice sheet out to a horizon approaching 700 km, monitoring a neutrino detection volume of order 10^6 km^3. The direction to the event is measured by time differences

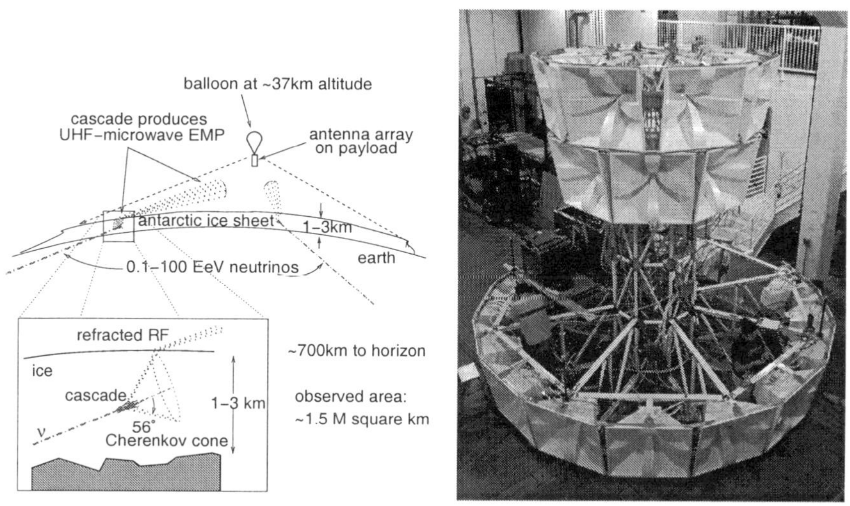

Fig. 1. (Left) Schematic of the ANITA concept, displaying the basic geometry for a detection of the coherent Cherenkov radio pulse generated by the cascades in Antarctic ice. (Right) The actual ANITA payload at its final setup

between antennas in the upper and lower clusters (Fig. 1 right). The statistical distribution of events should correlate with ice thickness, averaged over observable volume. As illustrated in Fig. 1, ANITA will search for radio pulses that arise from electromagnetic and hadronic cascades within the ice. The signals propagate through 1-3 km of ice with little attenuation. A radio pulse with zenith angle < 34° will refract through the air-ice interface and may be observed by ANITA. At ANITA energies, cascades initiated by electrons (e.g. in ν_e charged current events) are altered by LPM [12, 13] effects which narrows the width of the Cerenkov Cone pattern to considerably less than 5°. Cascades initiated by recoil hadrons are not affected, and provide the bulk of observable events for neutrino energies above 1 EeV [14].

3. ANITA-lite

A two-antenna prototype of ANITA, called ANITA-lite, was flown for 18 days on a Long-Duration Balloon (LDB). It was launched on December 17, 2003 as a piggyback instrument on board the Trans-Iron Galactic Element Recorder (TIGER) [15]. The ANITA-lite mission tested nearly every subsystem of ANITA, and monitored the Antarctic continent for impulsive Radio Frequency Interference (RFI) and ambient thermal noise levels as well as triggered events [16]. We discuss results that impact the upcoming ANITA experiment. Fig. 2 (left) shows the ANITA-lite and the projected ANITA aperture; not surprisingly, the ANITA effective volume improves more than one order of magnitude over the full range of energy of relevance. Fig. 2 (right) summarizes the science results obtained by ANITA-lite [17]. No excess events were observed for a UHE neutrinos search, thus a limit at 90% C.L. following E^{-2} spectrum was placed to $E_\nu^{-2}F < 1.6\times 10^{-6}$ GeV cm^{-2} s^{-1} sr^{-1} for $10^{18.5}$ eV < E_ν < $10^{23.5}$ eV. Various neutrino model pre-

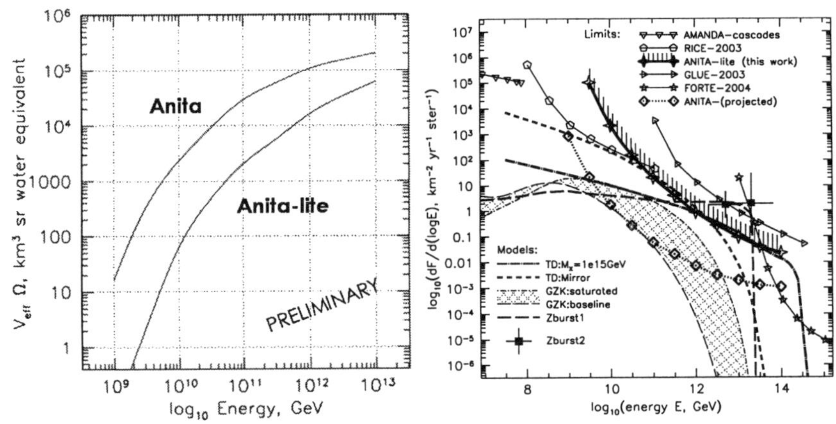

Fig. 2. (Left) Effective volume in terms of E_ν computed for ANITA and ANITA-lite. (Right) ANITA-lite experiential limit at 90% C.L. and projected ANITA sensitivity.

dictions have been considered, and the Z-burst model [18, 19, 20] has been excluded by this search. Assuming a 67% exposure over deep ice, typical of an LDB flight, we expect between 5-15 events from standard model GZK fluxes [21, 22], with the uncertainty primarily arising from assumptions about cosmic ray source evolution. A non-detection of these fluxes by ANITA would suggest non-standard phenomena in either particle physics or the astrophysics of cosmic ray propagation.

4. Radio attenuation in antarctic ice

The attenuation length of the ice beneath the South Pole Station was measured in 2004 [23]. A pair of broadband Ultra-High Frequency (UHF) horns, with range from 200 to 700 MHz, were used to make echograms by reflecting radar pulses off the bottom of the ice cap (depth 2810 m), and measuring the return amplitude in a separate receiver. Antarctic ice exhibits a horizontally layered structure, which creates small variations in the index of refraction. The near vertical penetration of the radio signal through the ice strata minimizes the impact of reflections due to variation of the index refraction. By making the assumption that the reflection coefficient off the

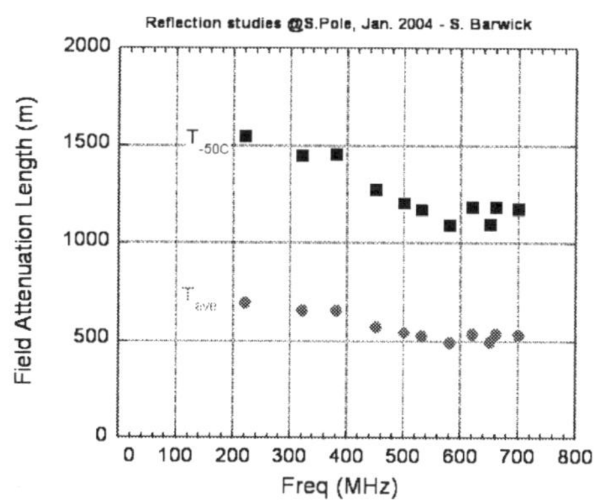

Fig. 3. Field attenuation length measured at the South Pole over the frequency range of relevance for the ANITA experiment.

bottom is unity, one can derive a lower limit on the attenuation length. Because of the logarithmic dependence of the derived attenuation length on most of the parameters, the uncertainty in the attenuation length is relatively small. A summary of the results is shown in Fig. 3, where the radio pulse propagates through ice which varies from $-50°C$ at the surface to near 0°C at the bottom. We correct the results to −45°C, a typical temperature for Antarctic ice. At this temperature the field attenuation length is ≥ 1 km between 200-700 MHz. These encouraging results define one of the most fundamental physical properties necessary for the success of ANITA.

5. ANITA probing physics beyond SM

From ANITA data is possible to infer a measurement of the neutrino cross-section at center-of-mass energies above 100 TeV, i.e. beyond the capabilities of current accelerator experiments. The method [24] consists of measuring the event rates of neutrinos emerging from the ice sheet and from the ice shelf, thus no accurate energy spectrum information is required, but good directional knowledge is desirable. Simulation studies show that the event

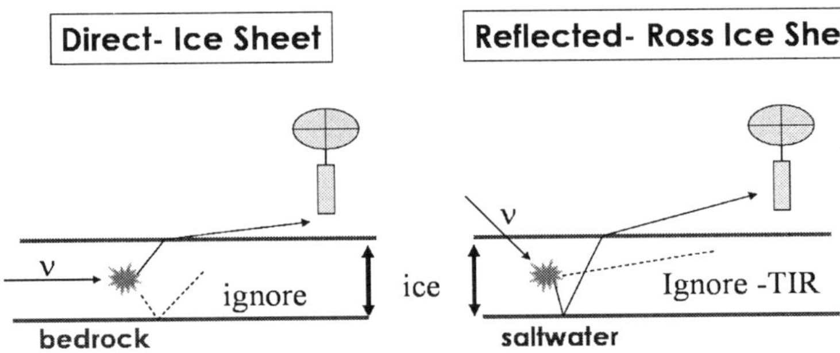

Fig. 4. (Left) Schematic concept for detecting direct rays, (right) and for observing reflected rays.

rate from the ice shelf depends linearly with the cross-section, while the rate from ice sheet is almost independent. Computing the ratio of the measured rates would yield a direct measurement of the neutrino cross-section. To accomplish this result the measurement of the event rates of the direct and reflected rays is required, as shown in Fig. 4. *Direct rays* are defined as the rays originated by almost horizontal neutrinos interacting within the ice sheet, which generate a upward going radio pulse. This pulse refracts at the ice-air boundary and can be detect by ANITA. While *reflected rays* are originated by downward going neutrinos which generate radio pulses that can reflect at the ice-water boundary with little attenuation, but don't have direct ray counterparts due to total internal reflection (TIR). Furthermore no reflected ray can be detected if the radio pulse propagated to the ice-bedrock boundary, due to the very poor reflection efficiency of the sedimen-

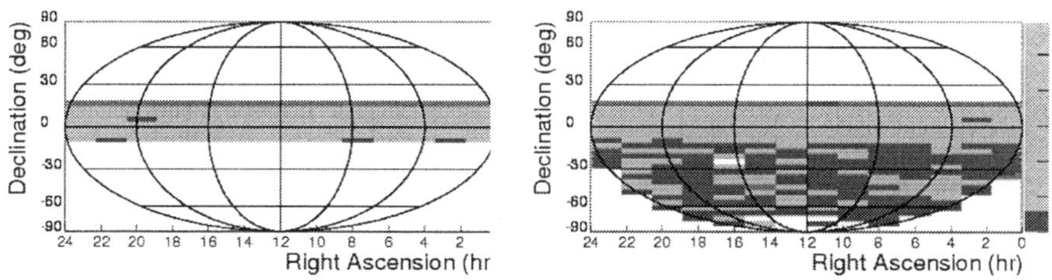

Fig. 5. (Left) ANITA sky survey for direct events only, (right) and for the combination of direct and reflected events.

tal material. Fig. 5 shows the observed sky by ANITA if only direct rays (left) or the combination of direct and reflected rays are detected (right). Direct rays are limited within few degrees with respect to the horizon because up-ward going neutrino events are absorbed by the Earth crust, and down-ward going events are limited by the TIR. This limitation is shown in Fig. 6 (left) which displays the angular distribution of direct events for energy E_ν^{20} eV and plotted for the σ_{SM} and $10\times\sigma_{SM}$. Reflected events, which are mainly generated by down-ward going events show a linear increase in rate correlated to the cross-section, because the neutrino pathlength is a small fraction of the interaction pathlength. Results from numerical simulation by taking into account the σ_{SM} calculation[25] and the differential GZK neutrino flux [21], are shown in Fig. 6 (right). For this calculation a $\sigma = 100\sigma_{SM}$ and $E_\nu = 10^{20}$eV were assumed. The reflected events cluster at smaller radial distances, where 90% of these events are originated from the ice shelves. Thus, the ratio of direct and reflected event rates can be inferred by the geographical location of the detected events. Fig. 7 summa-

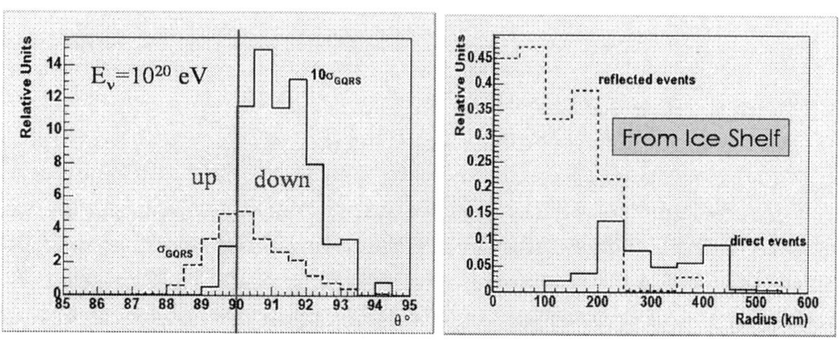

Fig. 6. (Left) Angular distribution of direct events for σ_{SM} and $10\times\sigma_{SM}$. (Right) Radial distribution of direct and reflected events.

rizes the results of this study, where the relative rate of direct and reflected events is plotted as a function of σ_ν normalized to σ_{SM} both for the GZK and the E^{-2} spectrum. As previously mentioned this simulation shows how the N_{dir} rate weakly depends on the σ_ν while the N_{ref} scales linearly with σ_ν except for very large cross-sections due to the effect of atmospheric absorption.

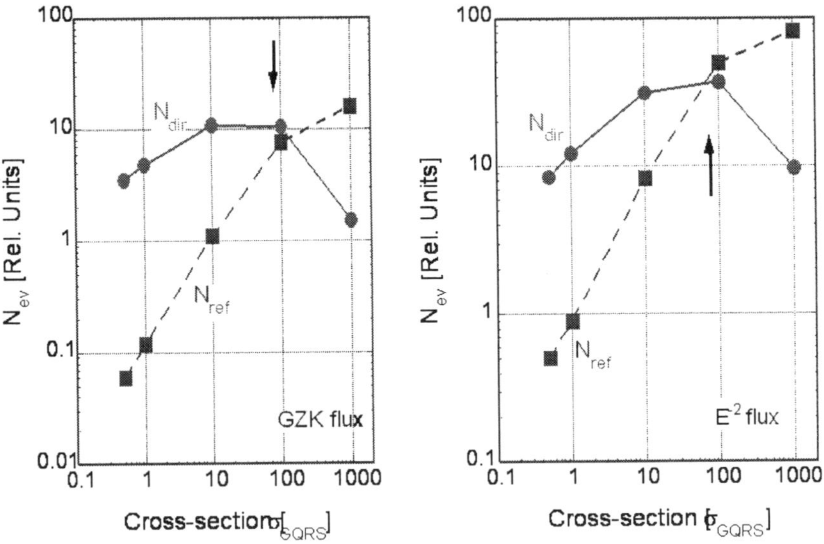

Fig. 7. N_{dir} and N_{ref} event rates as a function of σ_{SM} computed for GZK (left) and E^{-2} flux spectra (right)

6. ARIANNA

The Antarctic Ross Iceshelf ANtenna Neutrino Array is a complementary idea to the mission of the ANITA experiment. ANITA goals are targeting the detection of neutrinos with energy above $E_\nu \sim 10^{18}$ eV, whereas ARIANNA is designed to fill the energy gap by detecting neutrinos between $10^{17} - 3 \times 10^{18}$ eV (Fig. 9 left). The ARIANNA concept exploits two major advantages: the radio transparency of shelf ice, and the high reflectivity for radio signal of downward ν-events at the ice-water interface. The ARIANNA design (Fig. 8) consists of a 100×100 antenna array in a lattice formation with a 300 m separation. Four antennas, with features similar to the ANITA antennas, will form a station powered by solar panel device, and equipped with autonomous GPS clock and communication electronics borrowed by the experience of the Pierre Auger Observatory. The ice shelf in proximity to Minna Bluff, $\sim$ 150 km from McMurdo Station, has been identified as ideal location for the ARIANNA site. The high radio reflection efficiency and the quite radio environment of this location are particularly advantageous. However, the shelf ice is assumed to be $\sim$ 500

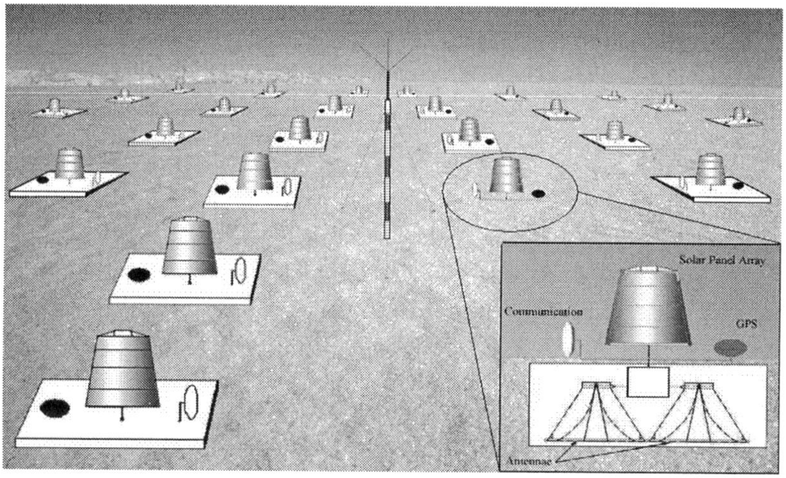

Fig. 8. Schematic of the ARIANNA concept, displaying the antenna array over the ice shelf, and the concept of a single station

m tick with a depth dependent temperature gradient, which impacts the attenuation length of the radio signals. Simulation studies show that the precise knowledge of the ice shelf attenuation length is more critical issue than the reflection efficiency. Therefore *in-situ* measurements of the radio attenuation length for frequency between 100MHz and 1GHz are planned to be performed during the 2006 campaign, and the installation of two prototype ARIANNA stations is also desirable. Fig. 9 (left) shows the results from these studies, where ARIANNA sensitivity for GZK neutrino detection improves by one order of magnitude compared to ANITA. ARIANNA should be able of detecting $\sim$ 40 events per 6 months lifetime. The energy threshold is lowered to $\sim 10^{17}$ eV, thanks the peculiar design of the detector array, and preliminary simulations show a relative energy resolution $\delta E/E \sim 1$, as well as an angular resolution of $\Delta\theta \sim 1^{\circ}$. Fig. 9 (right) shows the zenith-angular distribution for predicted and reconstructed events, as-

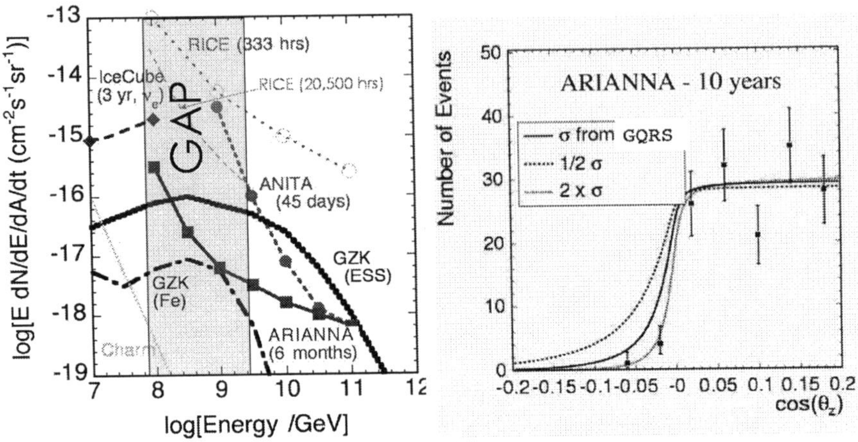

Fig. 9. (Left) Experimental flux limits and detector sensitivities of upcoming projects for the energy range above 10^{16} eV up to 10^{21} eV. (Right) Angular distribution of neutrino events observed over a 10 years of detector lifetime, the reconstructed events are compare to the $2 \times \sigma_{SM}$ prediction.

suming a $\sigma = 2 \times \sigma_{SM}$ and 10 years of detector operation. ARIANNA will have the unprecedented opportunity both to challenge numerous GZK model predictions, and to probe physics beyond the SM.

7. Conclusion

ANITA will use radio emission from the cascade induced by a neutrino interaction within the Antarctic ice cap to detect UHE neutrino interactions that occur within a million square km area. The remarkable transparency of Antarctic ice to radio waves makes this experiment possible, and the enormous volume of ice that can be simultaneously monitored leads to an unparalleled sensitivity to neutrinos in the energy range of 0.1 to 100 EeV. Remarkable science results were achieved with the successful 18.3-day flight of ANITA-lite. This is a very encouraging scenario for the ANITA mission, which will constraints GZK model predictions and provide insight to particle physics at the energy frontier. ARIANNA, a new idea based on combined experiences of the ANITA and the Pierre Auger projects, will measure with unprecedented sensitivity the GZK neutrino fluxes and constraints the neutrino cross section at the extreme high energy.

Acknowledgments

The author acknowledges support by the following agencies: NASA Research Opportunities in Space Science (ROSS) – The UH grant number is NASA NAG5-5387, Research Opportunities in Space Science – NSF and Raytheon Polar Services for Antarctic Support – NSBF for Balloon Launching and Operations – TIGER Collaboration for allowing ANITA-lite to fly as a piggyback – the Phi Beta Kappa Alumni in Southern California for providing travel grants – the Astrophysics Associates, Inc. – Italian Ministry of Education, European Physical Society – Ettore Majorana Foundation – Electron Tubes, Ltd. for providing the full scholarship at Erice.

References

1. F. Stecker, *Phys. Rev. Lett.* **21**, p. 1016 (1968).
2. V. Berezinsky and G. Zatsepin, *Phys. Lett.* **28B**, p. 453 (1969).
3. T. Stanev, These proceedings (2006).
4. K. Greisen, *Phys. Rev. Lett.* **16**, p. 748 (1966).
5. G. Zatsepin and V. Kuzmin, *JETP Letters* **4**, p. 78 (1966).
6. J. Feng and A. Shapere, *Phys. Rev. Lett.* **88**, p. 021303 (2002).
7. J. Alvarez-Muniz et al., *Phys. Rev. D* **65**, p. 124015 (2002).
8. G. Askaryan, *Sov. Phys. JETP* **14**, p. 441 (1962).
9. D. Saltzberg et al., *Phys. Rev. Lett.* **86**, p. 2802 (2001).
10. E. Zas et al., *Phys. Rev. D* **45**, 362 (1992).
11. A. Silvestri et al., Status of ANITA and ANITA-lite, in *Neutrinos and Explosive Events in the Universe*, Springer, New-York, **209**, 297-306, (2005).
12. L. D. Landau, *Dokl. Akad. Nauk Ser. Fiz.* **92**, 735 (1953).
13. A. B. Migdal, *Sov. Phys. JETP* **5**, p. 527 (1957).
14. J. Alvarez Muniz and E. Zas, *Phys. Lett. B* **434**, 396 (1998).
15. J. Link et al., Measurements of the ultra-heavy galactic cosmic-ray abundances between z=30 and z=40 with the tiger instrument, in *28th International Cosmic Ray Conference*, (2003).
16. P. Miocinovic et al., Tuning into uhe neutrinos in antarctica: The ANITA experiment, in *22nd Texas Symposium on Relativistic Astrophysics*, (2004).
17. S. Barwick et al., *Phys.Rev.Lett.* **96**, p. 171101 (2006).
18. B. Eberle et al., *Phys. Rev. D* **70**, p. 023007 (2004).
19. Z. Fodor et al., *Phys. Rev. Lett.* **88**, p. 171101 (2002).
20. T. Weiler, *Phys. Rev. Lett.* **49**, p. 234 (1982).
21. R. Engel et al., *Phys. Rev. D* **64**, p. 093010 (2001).
22. R. Protheroe and T. Stanev, *Phys. Rev. Lett.* **77**, p. 3708 (1996).
23. S. Barwick et al., *J. Glaciol.* **51** (2005).
24. S. Barwick, Using neutrino astronomy to probe physics at the highest energies, in *Neutrino Oscillations in Venice*, Venice, Italy, Edizioni Papergraf, 397-414 (2006).
25. R. Gandhi et al., *Phys. Rev. D* **58**, p. 093009 (1998).

LIST OF PARTICIPANTS

Aws Abdo	abdopa.msu.edu
Natalia Agafonova	agafonovalvd.ras.ru
Mohammed Said Ahrouaz	ahrouazlpnhep.in2p3.fr
Isabella Amore	amorelns.infn.it
Hendrik Bartko	Hbartkomppmu.mpg.de
Peter Biermann	plbiermannmpifr-bonn.mpg.de
Vadim Boyarkin	boyarkinlvd.ras.ru
Ronald Bruijn	rbruijnnikhef.nl
Emiliano Carmona	Carmonamppmu.mpg.de
Jeff Chancellor	Jeffery.c.chancellor1jsc.nasa.gov
Fabiana Cossavella	Fabiana.cossavellaik.fzk.de
Milena Dattoli	dattolito.infn.it
Lev Dorman	lidphysics.technion.ac.il
Ioana Dutan	Idutanmpifr-bonn.mpg.de
Manel Errando	errandoifae.es
Piero Galeotti	piero.galeottito.infn.it
Markus Garczarczyk	garczmppmu.mpg.de
Jordan Goodman	goodmanumdgrb.umd.edu
Alessio Gorgi	gorgito.infn.it
Masaaki Hayashida	mahayamppmu.mpg.de
Joerg Hörandel	Hoerandelik.fzk.de
Stefan Hoppe	Stefan.hoppempi-hd.mpg.de
Ching-Cheng Hsu	cchsumppmu.mpg.de
Scott Hughes	shugheshbar.wustl.edu
Giovanni Imponente	giovanni.imponentecern.ch
Clancy James	clancy.jamesstudent.adelaide.edu.au
Manasvita Joshi	joshihelios.phy.ohiou.edu
Sihem Kalli	sihemkhotmail.com
John Kelley	john.kelleyicecube.wisc.edu

Tomasz Lanczewski	tomasz.lanczewskiifj.edu.pl
Kuen Lee	kleehbar.wustl.edu
Peter Meszaros	pmeszarosastro.psu.edu
Jamal Mimouni	jamalmimounihotmail.com
Jelena Ninkovic	Ninkovicmppmu.mpg.de
Sven Over	overhep.physik.uni-siegen.de
Gabriela Pavalas	Gpavalasvenus.nipne.ro
Lech Wiktor Piotrowski	lewhoofuw.edu.pl
Agnieszka Pleti	abiszagbyk.oa.uj.edu.pl
Paolo Privitera	paolo.priviteraroma2.infn.it
Vladimir Ptuskin	vptuskinhotmail.com
Brian Rauch	briancosray.wustl.edu
Fabian Schuessler	Fabian.schuesslerik.fzk.de
David Seckel	seckelbartol.udel.edu
Kenji Shinozaki	kenjikrymppmu.mpg.de
Andrea Silvestri	silvestruci.edu
Arthur Smith	a.smith1physics.ox.ac.uk
Marcin Sokolowski	msokfuw.edu.pl
Christian Spiering	Csspierifh.de
Marko Stalevski	smarkobeotel.yu
Todor Stanev	stanevbartol.udel.edu
Jaroslaw Stasielak	stasielath.if.uj.edu.pl
Alessio Tamburro	alessio.tamburroik.fzk.de
Igor Telezhinsky	igor_tonline.com.ua
Omar Tibolla	omar.tibollapd.infn.it
Ralf Ulrich	ralf.ulrichik.fzk.de
Veronique VanElewyck	veroipno.in2p3.fr
Vlasios Vasileiou	vlasisvaumd.edu
Heinrich Völk	Heinrich.voelkmpi-hd.mpg.de
John P. Wefel	wefelphunds.phys.lsu.edu
Alan Wells	awstar.le.ac.uk